SpringerBriefs in Applied Sciences and Technology

Nanoscience and Nanotechnology

Series editor

Hilmi Volkan Demir, Nanyang Technological University, Singapore, Singapore

Nanoscience and nanotechnology offer means to assemble and study superstructures, composed of nanocomponents such as nanocrystals and biomolecules, exhibiting interesting unique properties. Also, nanoscience and nanotechnology enable ways to make and explore design-based artificial structures that do not exist in nature such as metamaterials and metasurfaces. Furthermore, nanoscience and nanotechnology allow us to make and understand tightly confined quasi-zero-dimensional to two-dimensional quantum structures such as nanoplatelets and graphene with unique electronic structures. For example, today by using a biomolecular linker, one can assemble crystalline nanoparticles and nanowires into complex surfaces or composite structures with new electronic and optical properties. The unique properties of these superstructures result from the chemical composition and physical arrangement of such nanocomponents (e.g., semiconductor nanocrystals, metal nanoparticles, and biomolecules). Interactions between these elements (donor and acceptor) may further enhance such properties of the resulting hybrid superstructures. One of the important mechanisms is excitonics (enabled through energy transfer of exciton-exciton coupling) and another one is plasmonics (enabled by plasmon-exciton coupling). Also, in such nanoengineered structures, the light-material interactions at the nanoscale can be modified and enhanced, giving rise to nanophotonic effects.

These emerging topics of energy transfer, plasmonics, metastructuring and the like have now reached a level of wide-scale use and popularity that they are no longer the topics of a specialist, but now span the interests of all "end-users" of the new findings in these topics including those parties in biology, medicine, materials science and engineerings. Many technical books and reports have been published on individual topics in the specialized fields, and the existing literature have been typically written in a specialized manner for those in the field of interest (e.g., for only the physicists, only the chemists, etc.). However, currently there is no brief series available, which covers these topics in a way uniting all fields of interest including physics, chemistry, material science, biology, medicine, engineering, and the others.

The proposed new series in "Nanoscience and Nanotechnology" uniquely supports this cross-sectional platform spanning all of these fields. The proposed briefs series is intended to target a diverse readership and to serve as an important reference for both the specialized and general audience. This is not possible to achieve under the series of an engineering field (for example, electrical engineering) or under the series of a technical field (for example, physics and applied physics), which would have been very intimidating for biologists, medical doctors, materials scientists, etc.

The Briefs in NANOSCIENCE AND NANOTECHNOLOGY thus offers a great potential by itself, which will be interesting both for the specialists and the non-specialists.

More information about this series at http://www.springer.com/series/11713

Pedro Ludwig Hernández Martínez

Alexander Govorov

Hilmi Volkan Demir

Understanding and Modeling Förster-type Resonance Energy Transfer (FRET)

FRET from Single Donor to Single Acceptor and Assemblies of Acceptors, Vol. 2

 Springer

Pedro Ludwig Hernández Martínez
School of Physical and Mathematical
Sciences, LUMINOUS! Centre of
Excellence for Semiconductor Lighting
and Displays, TPI—The Institute of
Photonics
Nanyang Technological University
Singapore
Singapore

Alexander Govorov
Department of Physics and Astronomy
Ohio University
Athens, OH
USA

Hilmi Volkan Demir
Department of Electrical and Electronics
Engineering, Department of Physics, and
UNAM—National Nanotechnology
Research Centre and Institute of Materials
Science and Nanotechnology
Bilkent University
Ankara
Turkey

and

School of Electrical and Electronic
Engineering, School of Physical and
Mathematical Sciences, LUMINOUS!
Centre of Excellence for Semiconductor
Lighting and Displays, TPI—The Institute
of Photonics
Nanyang Technological University
Singapore
Singapore

ISSN 2191-530X   ISSN 2191-5318   (electronic)
SpringerBriefs in Applied Sciences and Technology
ISSN 2196-1670   ISSN 2196-1689   (electronic)
Nanoscience and Nanotechnology
ISBN 978-981-10-1871-8   ISBN 978-981-10-1873-2   (eBook)
DOI 10.1007/978-981-10-1873-2

Library of Congress Control Number: 2016943801

Printed on acid-free paper

This Springer imprint is published by Springer Nature
The registered company is Springer Science+Business Media Singapore Pte Ltd.

Contents

Chapter 1
Applying Förster-Type Nonradiative Energy Transfer Formalism to Nanostructures with Various Directionalities: Dipole Electric Potential of Exciton and Dielectric Environment

In this chapter, we present analytical equations for the exciton electric potential inside and outside a nanostructure; including analytical expressions, for the long distance approximation, which are derived for the outside electric potential. Finally, the effective dielectric constant expressions, for this limit, are obtained. This chapter is reprinted (adapted) with permission from Ref. [1]. Copyright 2013 American Chemical Society.

1.1 Spherical Geometry: Nanoparticle Case

The electric potential for an exciton in the α-direction ($\alpha = x, y, z$), illustrated in Fig. 1.1a, is given by

$$\Phi_\alpha^{in} = \left(\frac{ed_{exc}}{\varepsilon_{NP}}\right)\frac{\hat{\alpha}\cdot\mathbf{r}}{r^3}\left(1 + \frac{2(\varepsilon_{NP} - \varepsilon_0)}{\varepsilon_{NP} + 2\varepsilon_0}\frac{r^3}{R_{NP}^3}\right) \tag{1.1}$$

$$\Phi_\alpha^{out} = \left(\frac{ed_{exc}}{\varepsilon_{NP}}\right)\left(\frac{3\varepsilon_{NP}}{\varepsilon_{NP} + 2\varepsilon_0}\right)\frac{\mathbf{r}\cdot\hat{\alpha}}{r^3} \tag{1.2}$$

where ε_{NP} and ε_0 are the nanoparticle (NP) and medium dielectric constants, respectively. The electric potential is the same in any direction because of the spherical symmetry of the NP. In the long distance approximation the outside electric potential can be written as

© The Author(s) 2017
P.L. Hernández Martínez et al., *Understanding and Modeling Förster-type Resonance Energy Transfer (FRET)*, Nanoscience and Nanotechnology, DOI 10.1007/978-981-10-1873-2_1

$$\Phi_\alpha^{out} = \left(\frac{ed_{exc}}{\varepsilon_{eff}}\right)\frac{\mathbf{r}\cdot\hat{\boldsymbol{\alpha}}}{r^3} \qquad (1.3)$$

where ε_{eff} is the effective dielectric constant given by

$$\varepsilon_{eff} = \frac{\varepsilon_{NP} + 2\varepsilon_0}{3} \qquad (1.4)$$

1.2 Cylindrical Geometry: Nanowire Case

In this case, the electric potential for an α-exciton $(\alpha = x, y, z)$, illustrated in Fig. 1.1a, is

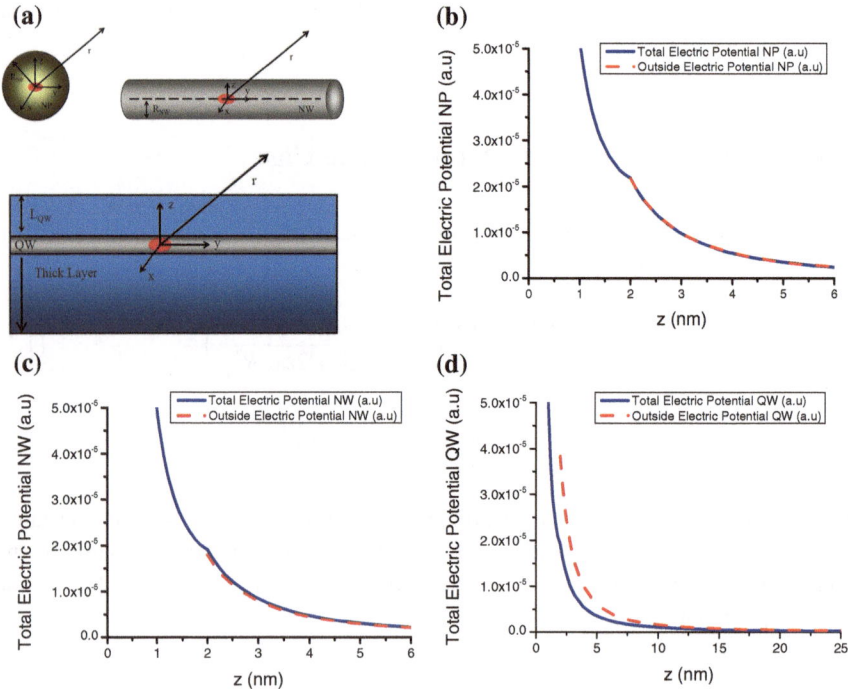

Fig. 1.1 a Schematic of an exciton in an NP, an NW, and a QW. *Red circle* represents an exciton in the α-direction. $R_{NP(NW)}$ is the NP (NW) radius. L_{QW} is the QW capping layer thickness. **b, c,** and **d** Electric potential along the "z" axis for a z-exciton. Total and long distance approximation electric potential for the z-exciton inside: **b** an NP; **c** an NW; and **d** a QW [Reprinted (adapted) with permission from Ref. [1] (Copyright 2013 American Chemical Society)]

$$\Phi_\alpha^{in} = \Phi_\alpha + \sum_m \int \left(e^{im\varphi} e^{-iky} A_m^\alpha(k) I_m(|k|\rho) \right) dk \tag{1.5}$$

$$\Phi_\alpha^{out} = \Phi_\alpha + \sum_m \int \left(e^{im\varphi} e^{-iky} B_m^\alpha(k) K_m(|k|\rho) \right) dk \tag{1.6}$$

where $I_m(|k|\rho)$ and $K_m(|k|\rho)$ are the modified Bessel functions of order m, and Φ_α is the α-exciton electric potential. After applying the boundary conditions at the surface of the nanowire (NW), the coefficients A_m^α and B_m^α are

$$A_m^\alpha(k) = \left(\frac{K_m(|k|R_{NW})}{I_m(|k|R_{NW})} \right) B_m^\alpha(k) \tag{1.7}$$

$$B_m^\alpha(k) = \frac{\frac{2}{|k|}(\varepsilon_0 - \varepsilon_{NW}) g_m^\alpha(|k|)}{\varepsilon_{NW} \left(\frac{K_m(|k|R_{NW})}{I_m(|k|R_{NW})} \right) I_m(|k|R_{NW}) + \varepsilon_0 K_m(|k|R_{NW})} \tag{1.8}$$

where $I_m(|k|R_{NW})$, $K_m(|k|R_{NW})$, and $g_m^\alpha(|k|)$ are defined as

$$I_m(|k|R_{NW}) = I_{m-1}(|k|R_{NW}) + I_{m+1}(|k|R_{NW}) \tag{1.9}$$

$$K_m(|k|R_{NW}) = K_{m-1}(|k|R_{NW}) + K_{m+1}(|k|R_{NW}) \tag{1.10}$$

$$g_m^\alpha(|k|) = \frac{1}{(2\pi)^2} \int_0^{2\pi} \int_{-\infty}^{\infty} d\varphi dy e^{-im\varphi} e^{iky} \left[\frac{\partial \Phi_\alpha}{\partial \rho} \right]_{\rho = R_{NW}} \tag{1.11}$$

For an exciton in the y-direction (along the cylinder axis), the coefficient B_m^y becomes

$$B_0^y(k) = \left(\frac{ed_{exc}}{\varepsilon_{NW}} \right) (\varepsilon_{NW} - \varepsilon_0) \frac{i}{\pi} |k| \left(\frac{1}{\left(\frac{K_0(|k|R_{NW}) I_1(|k|R_{NW})}{K_1(|k|R_{NW}) I_0(|k|R_{NW})} \right) \varepsilon_{NW} + \varepsilon_0} \right) \tag{1.12}$$

with an electric potential given by

$$\Phi_y^{out} = \left(\frac{ed_{exc}}{\varepsilon_{NW}} \right) \frac{y}{(\rho^2 + y^2)^{\frac{3}{2}}} + \int \left(e^{-iky} B_0^y(k) K_0(|k|\rho) \right) dk \tag{1.13}$$

In the long distance approximation, the coefficient B_m^y and the outside electric potential are simplified as

$$B_0^y(k) = \left(\frac{ed_{exc}}{\varepsilon_{NW}}\right)(\varepsilon_{NW} - \varepsilon_0)\frac{i}{\pi}|k|\left(\frac{1}{\varepsilon_0}\right) \tag{1.14}$$

$$\Phi_y^{out} = \left(\frac{ed_{exc}}{\varepsilon_{eff}}\right)\frac{y}{(\rho^2 + y^2)^{\frac{3}{2}}} \tag{1.15}$$

where ε_{eff} is the effective dielectric constant defined as

$$\varepsilon_{eff} = \varepsilon_0 \tag{1.16}$$

In the case of an exciton in the z-direction (perpendicular to the cylinder axis), the coefficient B_m^z which remains as B_1^z and B_{-1}^z, where B_1^z is given

$$B_1^z(k) = \left(\frac{ed_{exc}}{\varepsilon_{NW}}\right)(\varepsilon_0 - \varepsilon_{NW})\frac{|k|}{2\pi}\left(\frac{4}{3}\right)\frac{\frac{\left(\frac{1}{(|k|R_{NW})^2}G_{1,3}^{2,1}\left(\left(\frac{|k|R_{NW}}{2}\right)^2 \left|\begin{matrix} -\frac{1}{2} \\ 0,1,\frac{1}{2}\end{matrix}\right.\right) - K_2(|k|R_{NW})\right)}{(K_0(|k|R_{NW}) + K_2(|k|R_{NW}))}}{\frac{K_1(|k|R_{NW})}{I_1(|k|R_{NW})}\left(\frac{I_0(|k|R_{NW}) + I_2(|k|R_{NW})}{K_0(|k|R_{NW}) + K_2(|k|R_{NW})}\right)\varepsilon_{NW} + \varepsilon_0} \tag{1.17}$$

and $G_{1,3}^{2,1}\left(\left(\frac{|k|R_{NW}}{2}\right)^2 \left|\begin{matrix} -\frac{1}{2} \\ 0,1,\frac{1}{2}\end{matrix}\right.\right)$ is the Meijer G-function and $B_1^z = B_{-1}^z$. The electric potential is simplified as

$$\Phi_z^{out} = \left(\frac{ed_{exc}}{\varepsilon_{NW}}\right)\frac{\rho\cos(\varphi)}{(\rho^2 + y^2)^{\frac{3}{2}}} + 2\cos(\varphi)\int \left(e^{-iky}B_1^z(k)K_1(|k|\rho)\right)dk \tag{1.18}$$

In the long distance approximation, the coefficient B and the electric potential become

$$B_1^z(k) = -\left(\frac{ed_{exc}}{\varepsilon_{NW}}\right)(\varepsilon_0 - \varepsilon_{NW})\frac{1}{2\pi}|k|\left(\frac{1}{\varepsilon_{NW} + \varepsilon_0}\right) \tag{1.19}$$

$$\Phi_z^{out} = \left(\frac{ed_{exc}}{\varepsilon_{eff}}\right)\frac{\rho\cos(\varphi)}{(\rho^2 + y^2)^{\frac{3}{2}}} \tag{1.20}$$

where ε_{eff} is the effective dielectric constant defined as

$$\varepsilon_{eff} = \frac{\varepsilon_{NW} + \varepsilon_0}{2} \tag{1.21}$$

Similarly, for an exciton in the x-direction (perpendicular to the cylinder axis), the non-zero coefficients are B_1^x and B_{-1}^x, where B_1^x is given by

$$B_1^x(k) = \left(\frac{ed_{exc}}{\varepsilon_{NW}}\right)(\varepsilon_0 - \varepsilon_{NW})(-i)\frac{|k|}{2\pi}\left(\frac{4}{3}\right)\frac{\frac{1}{(|k|R_{NW})^2}G_{1,3}^{2,1}\left(\left(\frac{|k|R_{NW}}{2}\right)^2\left|\begin{matrix}-\frac{1}{2}\\0,1,\frac{1}{2}\end{matrix}\right.\right)-K_2(|k|R_{NW})}{(K_0(|k|R_{NW})+K_2(|k|R_{NW}))}\frac{K_1(|k|R_{NW})}{I_1(|k|R_{NW})}\left(\frac{I_0(|k|R_{NW})+I_2(|k|R_{NW})}{K_0(|k|R_{NW})+K_2(|k|R_{NW})}\right)\varepsilon_{NW}+\varepsilon_0 \tag{1.22}$$

with $B_{-1}^x = -B_1^x$ and the electric potential

$$\Phi_x^{out} = \left(\frac{ed_{exc}}{\varepsilon_{NW}}\right)\frac{\rho\sin(\varphi)}{(\rho^2+y^2)^{\frac{3}{2}}} + i2\sin(\varphi)\int\left(e^{-iky}B_1^x(k)K_1(|k|\rho)\right)dk \tag{1.23}$$

the coefficients B and the outside electric potential, in the long distance approximation, are simplified as

$$B_1^x(k) = \left(\frac{ed_{exc}}{\varepsilon_{NW}}\right)(\varepsilon_0 - \varepsilon_{NW})\frac{i}{2\pi}|k|\left(\frac{1}{\varepsilon_{NW}+\varepsilon_0}\right) \tag{1.24}$$

$$\Phi_x^{out} = \left(\frac{ed_{exc}}{\varepsilon_{eff}}\right)\frac{\rho\sin(\varphi)}{(\rho^2+y^2)^{\frac{3}{2}}} \tag{1.25}$$

where ε_{eff} is the effective dielectric constant, which is defined as

$$\varepsilon_{eff} = \frac{\varepsilon_{NW}+\varepsilon_0}{2} \tag{1.26}$$

1.3 Planar Geometry: Quantum Well Case

The electric potential, in cylindrical coordinates, for an α-exciton ($\alpha = x,y,z$), illustrated in Fig. 1.1a, is

$$\Phi_\alpha^{in} = \Phi_\alpha + \sum_m\int_0^\infty kdke^{-im\phi}J_m(k\rho)A_m^\alpha(k)\cosh(kz) \tag{1.27}$$

$$\Phi_\alpha^{out} = \Phi_\alpha + \sum_m\int_0^\infty kdke^{-im\phi}J_m(k\rho)B_m^\alpha(k)Exp(-k|z|) \tag{1.28}$$

where $J_m(k\rho)$ is the Bessel function of order m, and Φ_α is the α-exciton electric potential. After applying the boundary conditions at the surface of the QW, the coefficients A_m^α and B_m^α are

$$A_m^\alpha(k) = \left(\frac{\exp(-|k|\,L_{QW})}{\cosh(|k|\,L_{QW})}\right) B_m^\alpha(k) \tag{1.29}$$

$$B_m^\alpha(k) = \frac{(\varepsilon_0 - \varepsilon_{QW})h_m^\alpha(|k|)}{k(\varepsilon_{QW}\tanh(|k|\,L_{QW}) + \varepsilon_0)e^{-|k|\,L_{QW}}} \tag{1.30}$$

where $h_m^\alpha(|k|)$ is defined as

$$h_m^\alpha(|k|) = \frac{1}{(2\pi)}\int\limits_0^{2\pi}\int\limits_0^\infty d\varphi\rho d\rho e^{im\varphi}J_m(k\rho)\left[\frac{\partial\Phi_\alpha}{\partial z}\right]_{z=L_{QW}} \tag{1.31}$$

For an exciton in the z-direction, the non-zero coefficient is

$$B_0^z(k) = \left(\frac{ed_{exc}}{\varepsilon_{QW}}\right)\frac{(\varepsilon_{QW} - \varepsilon_0)}{(\varepsilon_{QW}\tanh(kL_{QW}) + \varepsilon_0)} \tag{1.32}$$

and the electric potential is

$$\Phi_z^{out} = \left(\frac{ed_{exc}}{\varepsilon_{QW}}\right)\frac{z}{(\rho^2 + z^2)^{\frac{3}{2}}} + \int\limits_0^\infty kdkJ_0(k\rho)B_0^z(k)Exp(-k|z|) \tag{1.33}$$

Thus, in the long distance approximation, the coefficient B and the electric potential are simplified as

$$B_0^z(k) \approx \left(\frac{ed_{exc}}{\varepsilon_{QW}}\right)\frac{(\varepsilon_{QW} - \varepsilon_0)}{\varepsilon_0} \tag{1.34}$$

$$\Phi_z^{out} = \left(\frac{ed_{exc}}{\varepsilon_{eff}}\right)\frac{z}{(\rho^2 + z^2)^{\frac{3}{2}}} \tag{1.35}$$

where ε_{eff} is the effective dielectric constant defined as

$$\varepsilon_{eff} = \varepsilon_0 \tag{1.36}$$

In the case of an exciton in the x-direction, the non-zero B coefficients are $B_1^x(k)$ and $B_{-1}^x(k)$, where $B_1^x(k) = B_{-1}^x(k)$ and

$$B_1^x(k) = \frac{1}{2}\left(\frac{ed_{exc}}{\varepsilon_{QW}}\right)\frac{(\varepsilon_{QW} - \varepsilon_0)}{(\varepsilon_{QW}\tanh(kL_{QW}) + \varepsilon_0)} \tag{1.37}$$

the outside electric potential is

$$\Phi_x^{out} = \left(\frac{ed_{exc}}{\varepsilon_{QW}}\right)\frac{\rho\cos(\phi)}{(\rho^2+z^2)^{\frac{3}{2}}} + 2\cos(\phi)\int\limits_0^\infty kdkJ_1(k\rho)B_1^x(k)Exp(-k|z|) \quad (1.38)$$

In the long distance approximation, the coefficient B and the electric potential are simplified into

$$B_0^x(k) \approx \frac{1}{2}\left(\frac{ed_{exc}}{\varepsilon_{QW}}\right)\frac{(\varepsilon_{QW}-\varepsilon_0)}{\varepsilon_0} \quad (1.39)$$

$$\Phi_x^{out} = \left(\frac{ed_{exc}}{\varepsilon_{eff}}\right)\frac{\rho\cos(\phi)}{(\rho^2+z^2)^{\frac{3}{2}}} \quad (1.40)$$

where ε_{eff} is the effective dielectric constant defined as

$$\varepsilon_{eff} = \varepsilon_0 \quad (1.41)$$

Similarly, for an exciton in the y-direction, the non-zero B coefficients are $B_{-1}^y(k)$ and $B_1^y(k)$, where $B_{-1}^y(k) = -B_1^y(k)$ and

$$B_1^y(k) = \frac{i}{2}\left(\frac{ed_{exc}}{\varepsilon_{QW}}\right)\frac{(\varepsilon_{QW}-\varepsilon_0)}{(\varepsilon_{QW}\tanh(kL_{QW})+\varepsilon_0)} \quad (1.42)$$

with the electric potential given by

$$\Phi_y^{out} = \left(\frac{ed_{exc}}{\varepsilon_{QW}}\right)\frac{\rho\sin(\phi)}{(\rho^2+z^2)^{\frac{3}{2}}} - i2\sin(\phi)\int\limits_0^\infty kdkJ_1(k\rho)B_1^y(k)Exp(-k|z|) \quad (1.43)$$

Thus, in the long distance approximation, the coefficient B and the outside electric potential are

$$B_1^y(k) \approx \frac{i}{2}\left(\frac{ed_{exc}}{\varepsilon_{QW}}\right)\frac{(\varepsilon_{QW}-\varepsilon_0)}{\varepsilon_0} \quad (1.44)$$

$$\Phi_y^{out} = \left(\frac{ed_{exc}}{\varepsilon_{eff}}\right)\frac{\rho\sin(\phi)}{(\rho^2+z^2)^{\frac{3}{2}}} \quad (1.45)$$

where ε_{eff} is the effective dielectric constant defined as

$$\varepsilon_{eff} = \varepsilon_0 \quad (1.46)$$

Table 1.1 Effective dielectric constant expressions for NP, NW, and QW cases in the long distance approximation

α-direction	NP	NW	QW
x	$\varepsilon_{eff} = \frac{\varepsilon_{NP} + 2\varepsilon_0}{3}$	$\varepsilon_{eff} = \frac{\varepsilon_{NW} + \varepsilon_0}{2}$	$\varepsilon_{eff} = \varepsilon_0$
y	$\varepsilon_{eff} = \frac{\varepsilon_{NP} + 2\varepsilon_0}{3}$	$\varepsilon_{eff} = \varepsilon_0$	$\varepsilon_{eff} = \varepsilon_0$
z	$\varepsilon_{eff} = \frac{\varepsilon_{NP} + 2\varepsilon_0}{3}$	$\varepsilon_{eff} = \frac{\varepsilon_{NW} + \varepsilon_0}{2}$	$\varepsilon_{eff} = \varepsilon_0$

This table follows the geometries given in Fig. 1.1 [Reprinted (adapted) with permission from Ref. [1] (Copyright 2013 American Chemical Society)]

A summary for the effective dielectric constant, for the long distance approximation, is given in Table 1.1. Table 1.1 shows the screening factor in the electric potential for different confinement geometries, which corresponds to the NP, NW, and QW cases. This screening factor comes from the boundaries conditions of the electric potential at the interface between the nanostructure (NP, NW, and QW) and the medium. For example, the screening factor for the NP case is the same for an exciton in the x-, y- and z-direction because of its spherical symmetry. In the cylindrical symmetry (NW case), an exciton in the cylindrical main axis does not have any screening factor. However, an exciton perpendicular to the cylindrical main axis has a screening factor as shown in Table 1.1. In the QW case, the screening factor is the same for the x-, y- and z-direction because the QW was considered infinitesimal thin. Table 1.1 follows the geometries sketched in Fig. 1.1a.

Figure 1.1 depicts the total and long distance approximation electric potentials for a z-exciton along the z axis. Figure 1.1b shows electric potentials in both the total and long distance approximation for a z-exciton inside an NP. It can be observed that both electric potentials overlap with each other because of the spherical symmetry of the NP nanostructure. The total and long distance approximation electric potentials for a z-exciton in a NW are depicted in Fig. 1.1c. In close proximity to the NW surface, the long distance approximation underestimates the exciton electric potential, as it is shown in Fig. 1.1c. In the QW case, the long distance approximation overestimates the exciton electric potential in the close proximity to the QW surface (Fig. 1.1d). This is an opposite effect compared to the NW case. These underestimation and overestimation of the electric potential, for NW and QW, respectively, is due to the fact that at short distances the long distance approximation do not apply and higher effects need to be considered. However, in all cases, at long distances the total electric potential converges into the long distance approximation (Fig. 1.1b–d).

Reference

1. P.L. Hernández-Martínez, A.O. Govorov, H.V. Demir, Generalized theory of Förster-type nonradiative energy transfer in nanostructures with mixed dimensionality. J. Phys. Chem. C **117**, 10203–10212 (2013)

Chapter 2
Förster-Type Nonradiative Energy Transfer Rates for Nanostructures with Various Dimensionalities

In this chapter, we derive the energy transfer rate for the cases of $X \rightarrow NP$ (nanoparticle), $X \rightarrow NW$ (nanowire), and $X \rightarrow QW$ (quantum well), where X is an NP, an NW, or a QW, and obtain simply expression for the long distance approximation. This chapter is reprinted (adapted) with permission from Ref. [1]. Copyright 2013 American Chemical Society.

We need to recall the results in Chap. 5 from Understanding and Modeling Förster-type Resonance Energy Transfer (FRET) Vol. 1, where the Fermi's Golden Rule is simplified into

$$\gamma_{trans} = \frac{2}{\hbar} \, \text{Im} \left[\int dV \left(\frac{\varepsilon_A(\omega)}{4\pi} \right) \mathbf{E}_{in}(\mathbf{r}) \cdot \mathbf{E}_{in}^*(\mathbf{r}) \right] \tag{2.1}$$

And

$$\mathbf{E}(\mathbf{r}) = -\nabla \Phi(\mathbf{r}) \tag{2.2}$$

with

$$\Phi_\alpha(\mathbf{r}) = \left(\frac{ed_{exc}}{\varepsilon_{eff_D}} \right) \frac{(\mathbf{r} - \mathbf{r}_0) \cdot \hat{\boldsymbol{\alpha}}}{|\mathbf{r} - \mathbf{r}_0|^3} \tag{2.3}$$

In addition to the previous results, we also recall the results obtained in the previous chapter (Chap. 1) regarding to the effective dielectric constant summarized in Table 2.1.

© The Author(s) 2017

P.L. Hernández Martínez et al., *Understanding and Modeling Förster-type Resonance Energy Transfer (FRET)*, Nanoscience and Nanotechnology, DOI 10.1007/978-981-10-1873-2_2

Table 2.1 Effective dielectric constant expressions for the cases of an NP, an NW, and a QW in the long distance approximation [Reprinted (adapted) with permission from Ref. [1] (Copyright 2013 American Chemical Society)]

α-direction	NP	NW	QW
x	$\varepsilon_{eff_D} = \frac{\varepsilon_{NP_D} + 2\varepsilon_0}{3}$	$\varepsilon_{eff_D} = \frac{\varepsilon_{NW} + \varepsilon_0}{2}$	$\varepsilon_{eff_D} = \varepsilon_0$
y	$\varepsilon_{eff_D} = \frac{\varepsilon_{NP_D} + 2\varepsilon_0}{3}$	$\varepsilon_{eff_D} = \varepsilon_0$	$\varepsilon_{eff_D} = \varepsilon_0$
z	$\varepsilon_{eff_D} = \frac{\varepsilon_{NP_D} + 2\varepsilon_0}{3}$	$\varepsilon_{eff_D} = \frac{\varepsilon_{NW} + \varepsilon_0}{2}$	$\varepsilon_{eff_D} = \varepsilon_0$

2.1 Cases of Förster-Type Energy Transfer to an Nanoparticle: NP → NP, NW → NP, and QW → NP

Here, we report analytical equations for FRET rate when the donor is an NP, an NW, or a QW and the acceptor is always an NP (Fig. 2.1). Furthermore, for the long distance approximation, we obtain simplified expressions for the transfer rate for all three cases (NP → NP, NW → NP, and QW → NP).

The exciton transfer rate (2.1), when the acceptor is an NP, is given by

$$\gamma_{\alpha,\,trans} = \frac{2}{\hbar} \, \mathrm{Im} \left[\int_{NP_A} dV \left(\frac{\varepsilon_{NP_A}(\omega)}{4\pi} \right) \mathbf{E}_{\alpha,\,in}(\mathbf{r}) \cdot \mathbf{E}^*_{\alpha,\,in}(\mathbf{r}) \right] \tag{2.4}$$

where ε_{NP_A} is the dielectric function of the acceptor and $\mathbf{E}_{\alpha,\,in}(\mathbf{r})$ is the induced electric field of an α-exciton ($\alpha = x, y, z$) in the donor. Assuming that the donor size is smaller than the separation distance between D and A and using the spherical symmetry of the acceptor, the total electric potential for the acceptor can be written as

$$\Phi_{\alpha}^{out}(r, \theta, \phi) = \Phi_{\alpha}(r, \theta, \phi) + \sum_{l,m} \frac{B_{l,m}^{\alpha}}{r^{l+1}} Y_{l,m}(\theta, \phi) \tag{2.5}$$

$$\Phi_{\alpha}^{in}(r, \theta, \phi) = \sum_{l,m} A_{l,m}^{\alpha} r^l Y_{l,m}(\theta, \phi) \tag{2.6}$$

where $\Phi_{\alpha}(r, \theta, \phi)$ is the electric potential of the exciton in the donor; $Y_{l,m}(\theta, \phi)$ are the spherical harmonics; and $A_{l,m}^{\alpha}$ and $B_{l,m}^{\alpha}$ are the coefficients determined by the boundary conditions. For the spherical case, the boundary conditions at the acceptor's surface ($r = R_{NP_A}$) are

$$\Phi_{\alpha}^{in}(r = R_{NP_A}, \theta, \phi) = \Phi_{\alpha}^{out}(r = R_{NP_A}, \theta, \phi) \tag{2.7}$$

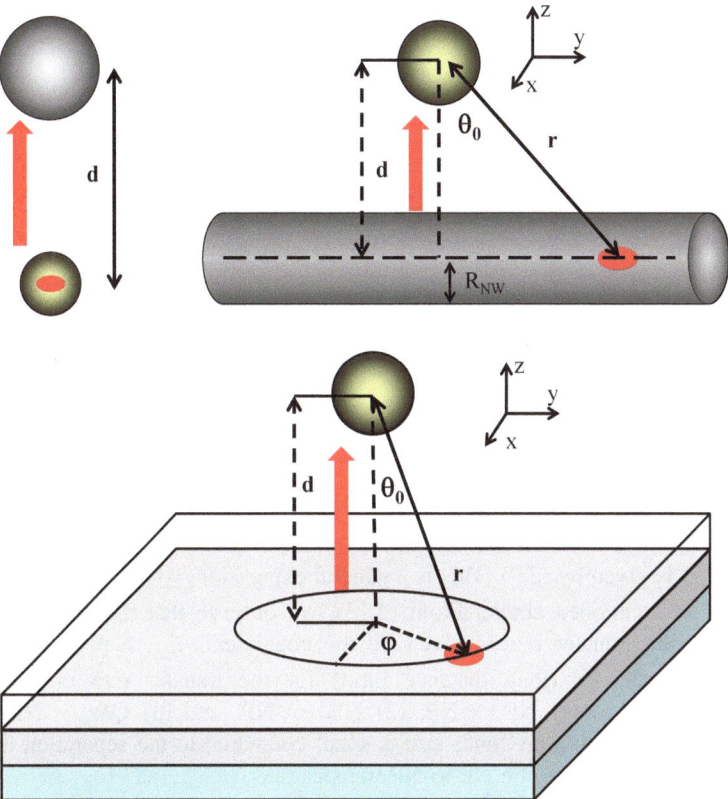

Fig. 2.1 Schematic for the energy transfer of NP → NP, NW → NP, and QW → NP. *Red arrows* show the energy transfer direction. *Red circles* represent an exciton in the α-direction. *d* is the separation distance. θ_0 is the azimuthal angle between *d* and *r*. *φ* is the radial angle [Reprinted (adapted) with permission from Ref. [1] (Copyright 2013 American Chemical Society)]

$$\varepsilon_{in}\left[\frac{\partial\Phi_\alpha^{in}(r,\theta,\phi)}{\partial r}\right]_{r=R_{NP_A}} = \varepsilon_{out}\left[\frac{\partial\Phi_\alpha^{out}(r,\theta,\phi)}{\partial r}\right]_{r=R_{NP_A}} \qquad (2.8)$$

where $\varepsilon_{in(out)}$ is the dielectric function inside (outside) the acceptor. Applying the boundary conditions (2.7) and (2.8) in (2.5) and (2.6), we obtain:

$$A_{l,m}^\alpha = \frac{B_{l,m}^\alpha}{R_{NP_A}^{2l+1}} + \frac{f_{l,m}^\alpha}{R_{NP_A}^l} \qquad (2.9)$$

$$B_{l,m}^\alpha = \frac{R_{NP_A}^{l+2}\left(\varepsilon_{out}g_{l,m}^\alpha - l\varepsilon_{in}\frac{f_{l,m}^\alpha}{R_{NP_A}}\right)}{l\varepsilon_{in} + (l+1)\varepsilon_{out}} \qquad (2.10)$$

with $f^{\alpha}_{l,m}$ and $g^{\alpha}_{l,m}$, which are given by

$$f^{\alpha}_{l,m} = \int_0^{2\pi} \int_0^{\pi} [\Phi_{\alpha}(r,\, \theta,\, \phi)]_{r=R_{NP_A}} Y^{*}_{l,m}(\theta,\, \phi) \sin(\theta)\, d\theta d\phi \qquad (2.11)$$

$$g^{\alpha}_{l,m} = \int_0^{2\pi} \int_0^{\pi} \left[\frac{\partial \Phi_{\alpha}(r,\, \theta,\, \phi)}{\partial r}\right]_{r=R_{NP_A}} Y^{*}_{l,m}(\theta,\, \phi) \sin(\theta)\, d\theta d\phi \qquad (2.12)$$

and $\varepsilon_{out} = \varepsilon_0$ is the dielectric constant of the medium, and $\varepsilon_{in} = \varepsilon_{NP_A}$ is the dielectric function of the acceptor. Combining (2.6) and (2.2) into (2.4), we obtain the energy transfer rate as

$$\gamma_{\alpha,\, trans} = \frac{2}{\hbar}\, \mathrm{Im}\left[\varepsilon_{NP_A}(\omega)\left(\frac{1}{4\pi}\right)\sum_{l,m}\left|A^{\alpha}_{l,m}\right|^2 \cdot l \cdot R^{2l+1}_{NP_A}\right] \qquad (2.13)$$

where $A^{\alpha}_{l,m}$ is given by (2.9). This is a general expression, which is valid under the assumption mentioned above. From (2.13), we observe that the distance dependency for the transfer rate is given by the coefficient $A^{\alpha}_{l,m}$. Now we derive an asymptotic behavior (long distance limit) for the transfer rate in the dipole approximation for: (1) NP $\rightarrow$ NP; (2) NW $\rightarrow$ NP; and (3) QW $\rightarrow$ NP. In all cases, we assume that the donor size is small compared to the separation distance d. Under this condition, the NP-to-NP transfer rate $(\gamma_{\alpha,\, trans})$ is

$$\gamma_{\alpha,\, trans} = \frac{2}{\hbar} b_{\alpha} \left(\frac{ed_{exc}}{\varepsilon_{eff_D}}\right)^2 \frac{R^3_{NP_A}}{d^6} \left|\frac{3\varepsilon_0}{\varepsilon_{NP_A}(\omega_{exc}) + 2\varepsilon_0}\right|^2 \mathrm{Im}\left[\varepsilon_{NP_A}(\omega_{exc})\right] \qquad (2.14)$$

where $b_{\alpha} = \frac{1}{3}, \frac{1}{3}, \frac{4}{3}$ for $\alpha = x, y, z$, respectively; d is the center-to-center distance between the donor and acceptor; and ε_{eff_D} the effective dielectric constant for the exciton in the donor, which is equal to $\varepsilon_{eff_D} = \frac{\varepsilon_{NP_D} + 2\varepsilon_0}{3}$ (Table 2.1) for the NP $\rightarrow$ NP case.

The transfer rate $(\gamma_{\alpha,\, trans})$ for the NW $\rightarrow$ NP is

$$\gamma_{\alpha,\, trans} = \frac{2}{\hbar} b_{\alpha} \left(\frac{ed_{exc}}{\varepsilon_{eff_D}}\right)^2 \frac{R^3_{NP_A}}{d^6} \cos^6(\theta_0) \left|\frac{3\varepsilon_0}{\varepsilon_{NP_A}(\omega_{exc}) + 2\varepsilon_0}\right|^2 \mathrm{Im}\left[\varepsilon_{NP_A}(\omega_{exc})\right] \quad (2.15)$$

where $b_{\alpha} = \frac{1}{3}, \frac{1}{3}, \frac{4}{3}$ for $\alpha = x, y, z$, respectively; θ_0 is the angle between d and $\mathbf{r}$; ε_{eff_D} the effective dielectric constant for the exciton in the donor, which is equal to

$\varepsilon_{eff_D} = \varepsilon_0$ for $\alpha = y$ (parallel to the cylindrical axis) and $\varepsilon_{eff_D} = \frac{\varepsilon_{NW} + \varepsilon_0}{2}$, $\alpha = x, z$ (perpendicular to the cylindrical axis) (Table 2.1).

Similarly, for the QW $\rightarrow$ NP, $\gamma_{\alpha, trans}$ is

$$\gamma_{\alpha, trans} = \frac{2}{\hbar} b_\alpha \left(\frac{ed_{exc}}{\varepsilon_{eff_D}} \right)^2 \frac{R_{NP_A}^3}{d^6} \cos^6(\theta_0) \left| \frac{3\varepsilon_0}{\varepsilon_{NP_A}(\omega_{exc}) + 2\varepsilon_0} \right|^2 \mathrm{Im} \left[\varepsilon_{NP_A}(\omega_{exc}) \right] \quad (2.16)$$

where $b_\alpha = \frac{1}{3}, \frac{1}{3}, \frac{4}{3}$ for $\alpha = x, y, z$, respectively; θ_0 is the angle between d and $\mathbf{r}$; and ε_{eff_D} the effective dielectric constant for the exciton in the donor, which is equal to $\varepsilon_{eff_D} = \varepsilon_0$ for $\alpha = x, y, z$ (Table 2.1).

The FRET rate for the NP $\rightarrow$ NP case follows the well-known asymptotic behavior $\gamma \propto d^{-6}$ [2]. Furthermore, the FRET rates are proportional to the imaginary part of the acceptor dielectric constant. Thus, an acceptor with strong absorption (large $\mathrm{Im} |\varepsilon_{NP_A}(\omega)|$) will have higher transfer rates. Moreover, in the cases of NW-to-NP and QW-to-NP, the transfer rate strongly depends on the distance and θ_0. In particular for the angle dependency, the main contribution comes from small θ_0 and decreases very fast as θ_0 increases. It is important to note that the transfer rate in these cases (NW-to-NP and QW-to-NP) follows the same distance dependency as the NP-to-NP transfer rate, which is $\gamma \propto d^{-6}$ [2]. These results suggest that the NRET rates are dictated by the acceptor's dimensionality, but not the donor's. It is worth mentioning that the FRET rate for the NW $\rightarrow$ NP and QW $\rightarrow$ NP cases have not been reported in early works. However, these missing cases for the FRET rates were reported in Ref. [1].

To illustrate the FRET rate, we present the average FRET rate in the long distance approximation as a function of the distance between CdTe D–A pair in Fig. 2.2. The acceptor dielectric function is taken from Ref. [3]. We assume that the acceptor exciton emission is at $\lambda = 582$ nm. In Fig. 2.2a, we consider the donor to be an NP, an NW, or a QW and the acceptor to be an NP. We set $\theta_0 = 0$ for the NP-to-NW and NP-to-QW cases. In this particular model, the larger average transfer rate is for the QW-to-NP case, and the smaller average transfer rate is for the NP-to-NP case. Figure 2.2c shows the energy transfer rate for the QW-to-NP case. Figure 2.2d depicts the contour profile plot for the QW-to-NP transfer rate. The top panel in Fig. 2.2d illustrates the energy transfer rate as a function of the distance at a fixed angle. Blue curve represents the case at $\theta_0 = 0$, and wine curve, at $\theta_0 = \pi/6$. The right panel in Fig. 2.2d shows the transfer rate as a function of the angle at a fixed distance. Red curve represents the case at $d = 3.3$ nm, and the green curve, at $d = 4.0$ nm. From Fig. 2.2c, d, the strong distance dependency of the transfer rate (2.15 and 2.16) is observed. Therefore, the main contribution for the energy transfer from a QW(NW) to an NP comes at short distances and small angles.

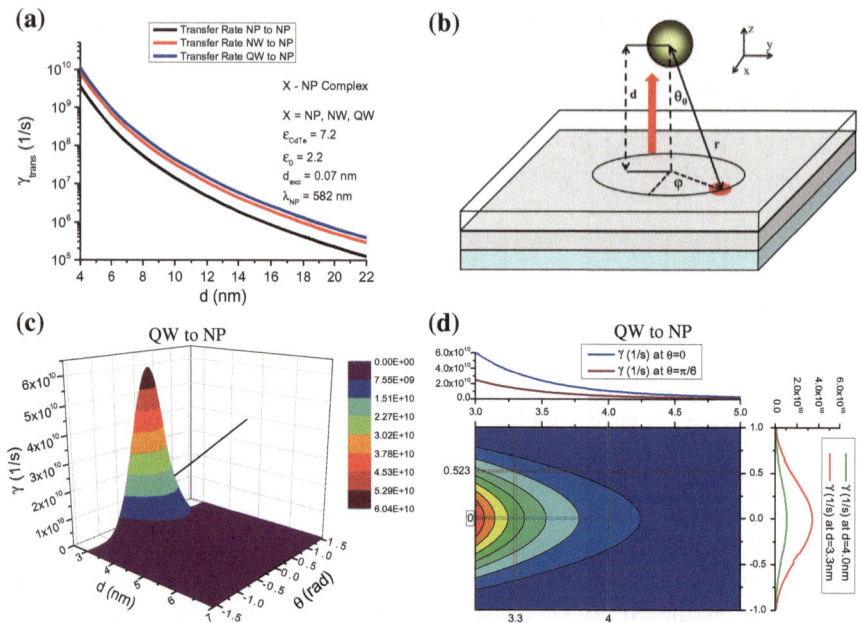

Fig. 2.2 **a** Average FRET rate for CdTe D–A pair. This shows the distance dependency of FRET rate for the NP → NP, NW → NP, and QW → NP cases. $\theta_0 = 0$ for the NW → NP and QW → NP pairs. **b** Schematic for the energy transfer of QW → NP case. **c** Average FRET rate for the CdTe D–A QW → NP pair as a function of the distance and angle. **d** Contour profile map for the average FRET rate for the CdTe D–A QW → NP pair, with the *top panel* at a fixed angle and *right panel* at a fixed distance [Reprinted (adapted) with permission from Ref. [1] (Copyright 2013 American Chemical Society)]

2.2 Cases of Förster-Type Energy Transfer to an Nanowire: NP → NW, NW → NW, and QW → NW

Here, we obtain analytical equations for the FRET rate when the donor is an NP, an NW, or a QW while the acceptor is always an NW (Fig. 2.3). We also obtained the simplified expressions for FRET rate in the long distance approximation for all these cases.

The transfer rate (2.1), when the acceptor is an NW, is written as

$$\gamma_{\alpha,\,trans} = \frac{2}{\hbar} \, \text{Im} \left[\int\limits_{NW_A} dV \left(\frac{\varepsilon_{NW_A}(\omega)}{4\pi} \right) \mathbf{E}_{\alpha,\,in}(\mathbf{r}) \cdot \mathbf{E}^*_{\alpha,\,in}(\mathbf{r}) \right] \qquad (2.17)$$

Here $\mathbf{E}_{\alpha,\,in}(\mathbf{r})$ is the induced electric field of an α-exciton ($\alpha = x, y, z$) in the donor and ε_{NW} is the dielectric function of the acceptor (NW). We assume that the

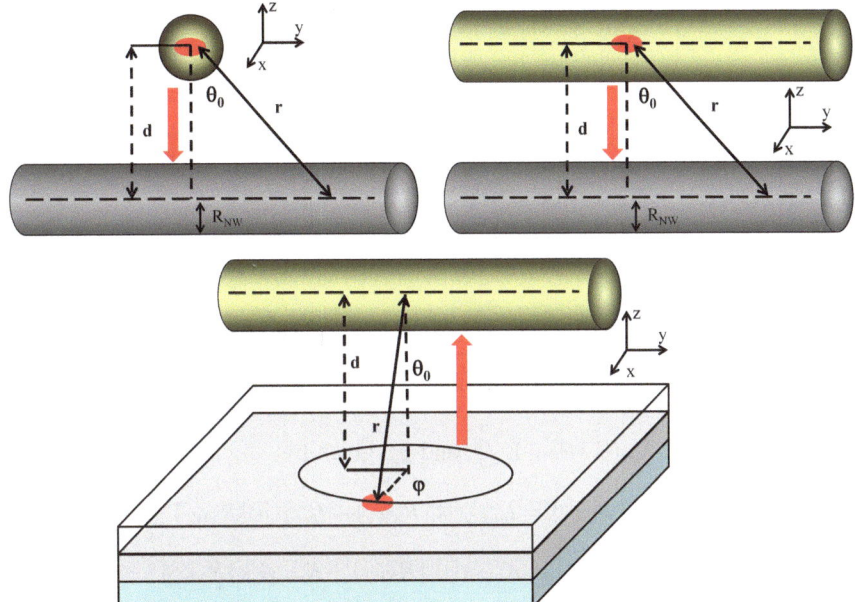

Fig. 2.3 Schematic for the energy transfer of NP → NW, NW → NW, and QW → NW. *Red arrows* show the energy transfer direction. *Red circles* represent an exciton in the α-direction. *d* is the separation distance. θ_0 is the azimuthal angle between *d* and *r*. φ is the radial angle [Reprinted (adapted) with permission from Ref. [1] (Copyright 2013 American Chemical Society)]

donor size is small compared to the D–A separation distance *d*. Taking advantage of the cylindrical symmetry of the acceptor, the total electric potential for the acceptor can be written as

$$\Phi_\alpha^{out}(\rho, \phi, z) = \Phi_\alpha(\rho, \phi, z) + \sum_m \int_{-\infty}^{\infty} dk e^{-ikz} B_m^\alpha(k) K_m(|k|\rho) e^{im\phi} \qquad (2.18)$$

$$\Phi_\alpha^{in}(\rho, \phi, z) = \sum_m \int_{-\infty}^{\infty} dk e^{-ikz} A_m^\alpha(k) I_m(|k|\rho) e^{im\phi} \qquad (2.19)$$

where $\Phi_\alpha(\rho, \phi, z)$ is the electric potential of the exciton in the donor; $I_m(|k|\rho)$ and $K_m(|k|\rho)$ are the modified Bessel functions; and $A_m^\alpha(k)$ and $B_m^\alpha(k)$ are the coefficients determined by the boundary conditions. For the cylindrical case, the boundary conditions at the acceptor's surface $(\rho = R_{NW_A})$ are

$$\Phi_\alpha^{in}(\rho = R_{NW_A}, \phi, z) = \Phi_\alpha^{out}(\rho = R_{NW_A}, \phi, z) \qquad (2.20)$$

$$\varepsilon_{in}\left[\frac{\partial\Phi_\alpha^{in}(\rho,\phi,z)}{\partial\rho}\right]_{\rho=R_{NW_A}} = \varepsilon_{out}\left[\frac{\partial\Phi_\alpha^{out}(\rho,\phi,z)}{\partial\rho}\right]_{\rho=R_{NW_A}} \tag{2.21}$$

where $\varepsilon_{in(out)}$ is the dielectric function inside (outside) the acceptor. Applying the boundary conditions (2.20) and (2.21) in (2.18) and (2.19), we arrive at

$$A_m^\alpha(k) = \frac{K_m(|k|R_{NW_A})}{I_m(|k|R_{NW_A})} B_m^\alpha(k) + \frac{f_m^\alpha(|k|)}{I_m(|k|R_{NW_A})} \tag{2.22}$$

$$B_m^\alpha(k) = \frac{\frac{2}{|k|}\varepsilon_{out}g_m^\alpha(|k|) - \varepsilon_{in}\frac{I_m(|k|R_{NW_A})}{I_m(|k|R_{NW_A})}f_m^\alpha(|k|)}{\varepsilon_{in}\left(\frac{I_m(|k|R_{NW_A})}{I_m(|k|R_{NW_A})}\right)K_m(|k|R_{NW_A}) + \varepsilon_{out}K_m(|k|R_{NW_A})} \tag{2.23}$$

with $I_m(|k|R_{NW_A})$, $K_m(|k|R_{NW_A})$, f_m^α, and g_m^α given by

$$I_m(|k|R_{NW_A}) = I_{m+1}(|k|R_{NW_A}) + I_{m-1}(|k|R_{NW_A}) \tag{2.24}$$

$$K_m(|k|R_{NW_A}) = K_{m+1}(|k|R_{NW_A}) + K_{m-1}(|k|R_{NW_A}) \tag{2.25}$$

$$f_m^\alpha = \frac{1}{(2\pi)^2}\int_0^{2\pi}\int_{-\infty}^{\infty}[\Phi_\alpha(\rho,\phi,z)]_{\rho=R_{NW_A}}e^{ikz}e^{-im\phi}\,dzd\phi \tag{2.26}$$

$$g_m^\alpha = \frac{1}{(2\pi)^2}\int_0^{2\pi}\int_{-\infty}^{\infty}\left[\frac{\partial\Phi_\alpha(\rho,\phi,z)}{\partial\rho}\right]_{\rho=R_{NW_A}}e^{ikz}e^{-im\phi}\,dzd\phi \tag{2.27}$$

and $\varepsilon_{out} = \varepsilon_0$ is the dielectric constant of the medium, and $\varepsilon_{in} = \varepsilon_{NW_A}$ is the dielectric function of the NW. Combining (2.19) and (2.2) into (2.17), we obtain that the energy transfer rate of

$$\gamma_{\alpha,trans} = \frac{2}{\hbar}\text{Im}\left[\frac{\varepsilon_{NW_A}(\omega_{exc})}{4\pi}\right](2\pi)^2\sum_m\int_{-\infty}^{\infty}dk|A_m^\alpha(|k|)|^2$$

$$\times\left(\frac{|k|^2}{4}\int_0^{R_{NW_A}}|I_m(|k|\rho)|^2\rho d\rho + m^2\int_0^{R_{NW_A}}|I_m(|k|\rho)|^2\frac{1}{\rho}d\rho + |k|^2\int_0^{R_{NW_A}}|I_m(|k|\rho)|^2\rho d\rho\right) \tag{2.28}$$

where $A_m^\alpha(k)$ is given by (2.22). This is a general expression, which is valid under the mentioned assumptions. Note that the distance dependency of FRET rate is given by the coefficient $A_m^\alpha(k)$. For the long distance approximation, we derive the transfer rate equations for the NP-to-NW, NW-to-NW and QW-to-NW cases. Thus, the transfer rate is

$$\gamma_{\alpha,trans} = \frac{2}{\hbar}\left(\frac{ed_{exc}}{\varepsilon_{effD}}\right)^2\left(\frac{3\pi}{32}\right)\frac{R_{NW_A}^2}{d^5}\left(a_\alpha + b_\alpha\left|\frac{2\varepsilon_0}{\varepsilon_{NW_A}(\omega_{exc})+\varepsilon_0}\right|^2\right)\text{Im}\left[\varepsilon_{NW_A}(\omega_{exc})\right]$$

(2.29)

where $a_\alpha = 0, \frac{9}{16}, \frac{15}{16}$; $b_\alpha = 1, \frac{15}{16}, \frac{41}{16}$ for $\alpha = x, y, z$, respectively; d is the center-to-center distance between the donor and the acceptor; and ε_{effD} is the effective dielectric constant for the exciton in the donor, which is equal to $\varepsilon_{effD} = \frac{\varepsilon_{NP_D}+2\varepsilon_0}{3}$ for NP $\rightarrow$ NW. In the NW $\rightarrow$ NW case, the effective dielectric constant is $\varepsilon_{effD} = \varepsilon_0$ for $\alpha = y$ (parallel to the cylindrical axis) and $\varepsilon_{effD} = \frac{\varepsilon_{NW_D}+\varepsilon_0}{2}$ for $\alpha = x, z$ (perpendicular to the cylindrical axis) (Table 2.1). Likewise, the QW-to-NW transfer rate ($\gamma_{\alpha,trans}$) is given by

$$\gamma_{\alpha,trans} = \frac{2}{\hbar}\left(\frac{ed_{exc}}{\varepsilon_{effD}}\right)^2\left(\frac{3\pi}{32}\right)\frac{R_{NW_A}^2}{d^5}\cos^5(\theta_0)\left(a_\alpha + b_\alpha\left|\frac{2\varepsilon_0}{\varepsilon_{NW_A}(\omega_{exc})+\varepsilon_0}\right|^2\right)\text{Im}\left[\varepsilon_{NW_A}(\omega_{exc})\right]$$

(2.30)

where θ_0 is the angle between d and $\mathbf{r}$ and ε_{effD} is the effective dielectric constant for the exciton in the donor, which is equal to $\varepsilon_{effD} = \varepsilon_0$ for $\alpha = x, y, z$ (Table 2.1).

As expected, the asymptotic behavior for the NRET rate of the QW $\rightarrow$ NW case follows $\gamma \propto d^{-5}$ [1]. This result is similar to the NP-to-NW and NW-to-NW cases, as reported in Refs. [4, 5], respectively. Similar to the previous section, the FRET rates strongly depend on the distance and θ_0, and a similar analysis can be made. Figure 2.4a depicts the average FRET rate for a CdTe D–A pair as a function of the distance, when the donor is an NP, an NW, or a QW while the acceptor is an NW in all cases. We set $\theta_0 = 0$ for the QW-to-NW case. We assume that the acceptor exciton emission is at $\lambda = 610$ nm and the acceptor dielectric function is taken from Ref. [3]. Note that the higher transfer rate is for the QW-to-NW, and the lower rate is for the NP-to-NW. Figure 2.4c, d depict the average FRET rate for a CdTe D–A pair as a function of the distance and θ_0, when the donor is a QW and the acceptor is an NW. Figure 2.4d shows the contour profile map for the QW-to-NW transfer rate. The top panel in Fig. 2.4d illustrates the energy transfer rate as a function of the distance at a fixed angle. Blue curve represents the case at $\theta_0 = 0$, and wine curve, at $\theta_0 = \pi/6$. The right panel in Fig. 2.4d shows the transfer rate as a function of the angle at a fixed distance. Red curve represents the behavior at $d = 3.3$ nm, and the green curve, at $d = 4.0$ nm. From Fig. 2.4a, c, d, show the strong distance dependency of the transfer rate (2.29, 2.30). Similar to the previous section, the main contribution for the energy transfer from a QW to an NW comes at short distances and small angles.

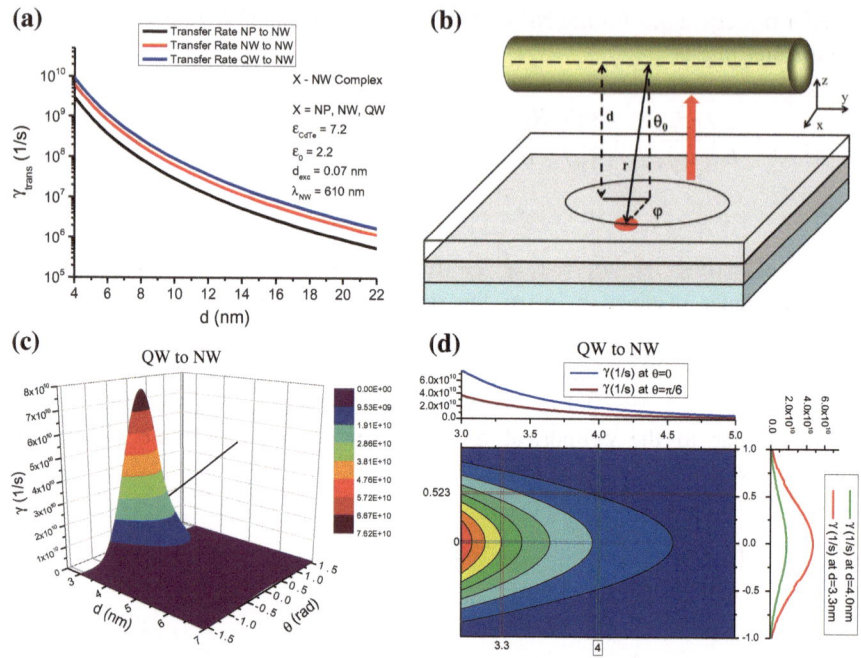

Fig. 2.4 **a** Average FRET rate for CdTe D–A pair. This plot illustrates the FRET distance dependency for NP $\rightarrow$ NW, NW $\rightarrow$ NW, and QW $\rightarrow$ NW cases. $\theta_0 = 0$ for QW to NW pairs. **b** Schematic for the energy transfer of QW $\rightarrow$ NW case. **c** Average FRET rate for the CdTe D–A QW $\rightarrow$ NW pair as a function of the distance and angle. **d** Contour profile map for the average FRET rate of QW $\rightarrow$ NW, with the *top panel* at a fixed angle, and the *right panel* at a fixed distance [Reprinted (adapted) with permission from Ref. [1] (Copyright 2013 American Chemical Society)]

2.3 Cases of Förster-Type Energy Transfer to a Quantum Well: NP $\rightarrow$ QW, NW $\rightarrow$ QW, and QW $\rightarrow$ QW

In this section, we obtain analytical equations for the FRET rate when the donor is an NP, an NW, or a QW while the acceptor is always a QW (Fig. 2.5). Moreover, the simplified expression for the FRET rate in the long distance approximation is obtained for all these cases.

The transfer rate (2.1), when the acceptor is a QW, is written as

$$\gamma_{\alpha,\,trans} = \frac{2}{\hbar} \, \mathrm{Im} \left[\int_{QW_A} dV \left(\frac{\varepsilon_{QW_A}(\omega)}{4\pi} \right) \mathbf{E}_{\alpha,\,in}(\mathbf{r}) \cdot \mathbf{E}^*_{\alpha,\,in}(\mathbf{r}) \right] \qquad (2.31)$$

where $\mathbf{E}_{\alpha,\,in}(\mathbf{r})$ represents the electric field of an α-exciton ($\alpha = x,\, y,\, z$) in the donor and ε_{QW_A} is the dielectric function of the acceptor (QW). Now we assume that the

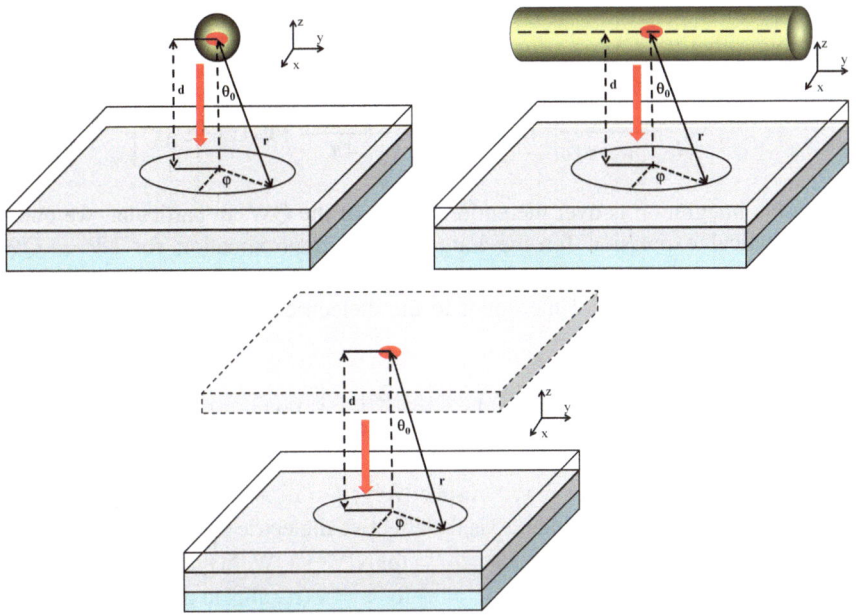

Fig. 2.5 Schematic for the energy transfer of NP → QW, NW → QW, and QW → QW. *Red arrows* show the energy transfer direction. *Red circles* represent an exciton in the α-direction. *d* is the separation distance. θ_0 is the azimuthal angle between *d* and *r*. φ is the radial angle [Reprinted (adapted) with permission from Ref. [1] (Copyright 2013 American Chemical Society)]

donor size is small compared to the D–A separation distance *d*. Furthermore, we consider a symmetric structure, consisting of a semiconductor QW of thickness L_w between two barriers of dielectric function ε_{QW_A}. One barrier has a film thickness L_l, while the other barrier is considered to be very thick (where we assume that this barrier is semi-infinite). The donor nanostructure is placed in front of the barrier with thickness L_l and we solve the problem for the case where the QW is very thin ($L_w \ll L_l$). Under these assumptions, the electric potential inside the barrier is

$$\Phi_{in}(\mathbf{r}) = \left(\frac{2\varepsilon_0}{\varepsilon_{QW_A} + \varepsilon_0}\right) \Phi_\alpha(\mathbf{r}) \tag{2.32}$$

where ε_0 is the dielectric constant of the matrix (surrounding the medium around the donor); ε_{QW_A} is the dielectric function of the barrier; and Φ_α is the electric potential of an α-exciton in the donor nanostructure. Combining (2.32) and (2.2) into (2.31), we obtain

$$\gamma_{\alpha, trans} = \frac{2}{\hbar} \left| \frac{2\varepsilon_0}{\varepsilon_{QW_A} + \varepsilon_0} \right|^2 \text{Im} \left[\int dV \left(\frac{\varepsilon_{QW_A}(\omega)}{4\pi} \right) \mathbf{E}_\alpha(\mathbf{r}) \cdot \mathbf{E}_\alpha^*(\mathbf{r}) \right] \tag{2.33}$$

where $E_\alpha(\mathbf{r})$ is the electric field created by an α-exciton in the donor. By using the assumption that the QW is very thin ($L_w \ll L_l$), the energy transfer rate becomes

$$\gamma_{\alpha, trans} = \frac{2}{\hbar} \left| \frac{2\varepsilon_0}{\varepsilon_{QW_A} + \varepsilon_0} \right|^2 \mathrm{Im} \left[\int_{QW_A} dS \left(\frac{\varepsilon_{QW_A}(\omega)}{4\pi} \right) E_\alpha(\mathbf{r}) \cdot \mathbf{E}_\alpha^*(\mathbf{r}) \right] \quad (2.34)$$

where the integration is over the entire surface of the QW. In particular, we obtain the analytical expression for the long distance approximation for NP $\to$ QW, NW $\to$ QW, and QW $\to$ QW. In all cases, we assume $d_b \gg L_W$ where d_b is the distance from the center of the donor to the dielectric barrier. Under these conditions, $\gamma_{\alpha, trans}$ becomes

$$\gamma_{\alpha, trans} = \frac{2}{\hbar} b_\alpha \left(\frac{e d_{exc}}{\varepsilon_{eff_D}} \right)^2 \frac{1}{d^4} \left| \frac{2\varepsilon_0}{\varepsilon_{QW_A} + \varepsilon_0} \right|^2 \mathrm{Im} \left[\varepsilon_{QW_A}(\omega_{exc}) \right] \quad (2.35)$$

where $b_\alpha = \frac{3}{16}, \frac{3}{16}, \frac{3}{8}$ for $\alpha = x, y, z$, respectively; $d = d_b + L_l$ is the distance between the donor and the acceptor; and ε_{eff_D} is the effective dielectric constant for the exciton in the donor, which is equal to $\varepsilon_{eff_D} = \frac{\varepsilon_{NP_D} + 2\varepsilon_0}{3}$ for NP $\to$ QW. In the NW $\to$ QW case, the effective dielectric constant is $\varepsilon_{eff_D} = \varepsilon_0$ for $\alpha = y$ (parallel to the cylindrical axis) and $\varepsilon_{eff_D} = \frac{\varepsilon_{NW} + \varepsilon_0}{2}$ for $\alpha = x, z$ (perpendicular to the cylindrical axis). For QW$\to$QW, $\varepsilon_{eff_D} = \varepsilon_0$ for $\alpha = x, y, z$ (Table 2.1). Note that the FRET rate for the NP $\to$ QW and QW $\to$ QW cases follow the well-known asymptotic behavior $\gamma \propto d^{-4}$ [6] and $\gamma \propto d^{-4}$ [7], respectively. Akin to the previous cases for the FRET rate, we have included the FRET rate for the NW $\to$ QW case, which was studied in Ref. [1].

Figure 2.6 shows the average FRET rate for a CdTe D–A pair as a function of the distance, when the donor is an NP, an NW, or a QW with the acceptor being a

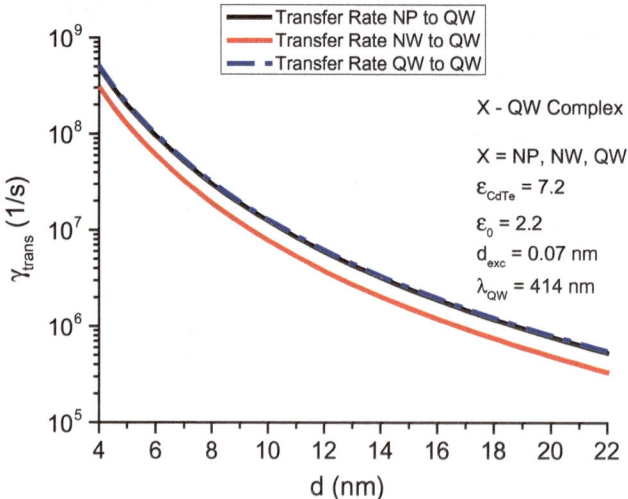

Fig. 2.6 Average FRET rate for a CdTe D–A pair. This plot shows the distance dependency of the FRET rate for the NP $\to$ QW, NW $\to$ QW, and QW $\to$ QW cases [Reprinted (adapted) with permission from Ref. [1] (Copyright 2013 American Chemical Society)]

QW in all cases. In this plot, we made similar assumptions as the previous section. Here, the faster transfer rate is for the QW → QW pair which slightly faster than the NP → QW pair; On the other hand, the lower transfer rate is for the NW → QW pair.

2.4 Example: Energy Transfer Between Nanoparticles and Nanowires

As an example we calculate Förster energy transfer from an optically excited NP to NW as shown in Fig. 2.7 [4]. The center-to-center distance between NP and NW is denoted as d, and the distance between the NP center and the NW surface is given by Δ. A NW, NP, and matrix are described with local dielectric constants denoted as ε_{NW}, ε_{NP}, and ε_0, respectively. The local dielectric constant approach provides us with a reliable description if the transferred exciton energy when the bandgap of a donor nanocrystal is not very close to the bandgap of a NW (acceptor). From (2.28), the transfer rate takes the form Ref. [4]

$$
\gamma_\alpha = \frac{2}{\hbar} \, \mathrm{Im} \left[\frac{\varepsilon_{NW}}{2\pi} \right] \cdot (2\pi)^2 \sum_m \int_{-\infty}^{\infty} dk \, | A^*(m,k) |^2
$$

$$
\times \left\{ \frac{k^2}{4} \int_0^{R_{NW}} \rho d\rho \, | I_{m+1}(k\rho) + I_{m-1}(k\rho) |^2 + m^2 \int_0^{R_{NW}} \frac{1}{\rho} d\rho \, | I_m(k\rho) |^2 + k^2 \int_0^{R_{NW}} \rho d\rho \, | I_m(k\rho) |^2 \right\}
$$

$$(2.36)$$

For the case where $d \gg R_{NW}$, we expand (2.36) in terms of the parameter R_{NW}/d and obtain a convenient relation:

$$
\gamma_\alpha(\omega_{exc}) = \frac{2}{\hbar} \frac{R_{NW}^2}{d^5} \left(\frac{ed_{exc}}{\varepsilon_{eff}} \right)^2 \frac{3\pi}{32} \left(a_\alpha + \left| \frac{\varepsilon_0}{\varepsilon_{NW} + \varepsilon_0} \right|^2 b_\alpha \right) \mathrm{Im} \, \varepsilon_{NW} \qquad (2.37)
$$

where the coefficient a_α is 15/16, 0, and 9/16 for $\alpha = x, y$ and z, respectively; the corresponding values for the coefficient b_α are: 41/4, 4, and 15/4. Notice that the

Fig. 2.7 Schematics of the coupled NP-NW system [Reprinted (abstract/excerpt/figure) with permission from Ref. [4] (Copyright 2008 by the American Physical Society)]

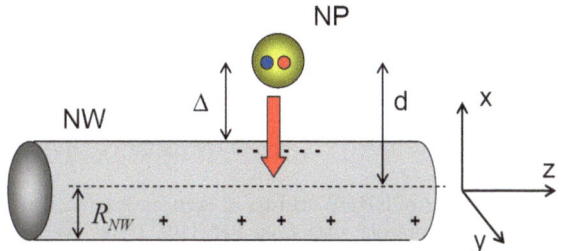

distance dependence of Förster transfer for the dipole-to-nanowire case is $\gamma_\alpha \propto 1/d^5$ as compared to the case of traditional dipole-dipole transfer $\gamma_{dipole-dipole} \propto 1/d^6$ [8, 9]. Slower spatial decay of the energy transfer rate comes from the one-dimensional character of a NW. Equation (2.36) is rather complicated, therefore (2.37) can be very convenient to estimate transfer rates in structures where $(R_{NW}/d) < 1$.

To illustrate the validity of (2.37), we numerically calculate the transfer rate for the cases of (1) CdTe NP–CdTe NW and (2) CdTe NP–Carbon Nanotube (CNT). In Fig. 2.8 shows the results for these complexes. The CdTe NPs and CdTe NWs was assembled and optically characterized in Ref. [10]. Experimental values for the FRET rates for NPs to NWs case were extracted from the photoluminescence spectra recorded during the assembly process. The experiment in Ref. [10] was performed with orange- and green-emitting CdTe NPs: $\lambda_{exc,orange\,NP} = 582$ nm ($R_{orange\,NP} = 2$ nm) and $\lambda_{exc,green\,NP} = 526$ nm ($R_{green\,NP} = 1.6$ nm). The NW radius $R_{NW} = 3.3$ nm and its emission is at $\lambda_{exc,NW} = 689$ nm. The NP-NW complex was assembled using the biotin-streptavidin biolinker with a length of 5 nm. The resultant NP-NW distances were estimated as: $d_{orange\,NP} = 10.3$ nm and $d_{green\,NP} = 9.9$ nm, with an estimated dipole moment of $d_{exc} \sim 0.08 nm$. From the experiment, it was determined that $\gamma_{trans,\,orange} = 1/16\,\text{ns}^{-1}$ and $\gamma_{trans,\,green} = 1/12\,\text{ns}^{-1}$ whereas the corresponding theoretical estimated values are: $\gamma_{trans,\,orange}^{theory} \approx 1/13.1\,\text{ns}^{-1}$ and $\gamma_{trans,\,green}^{theory} \approx 1/9\,\text{ns}^{-1}$. From here, we can say that the calculations provide us with reliable estimates for the FRET rates for NP→NW system (Figs. 2.7

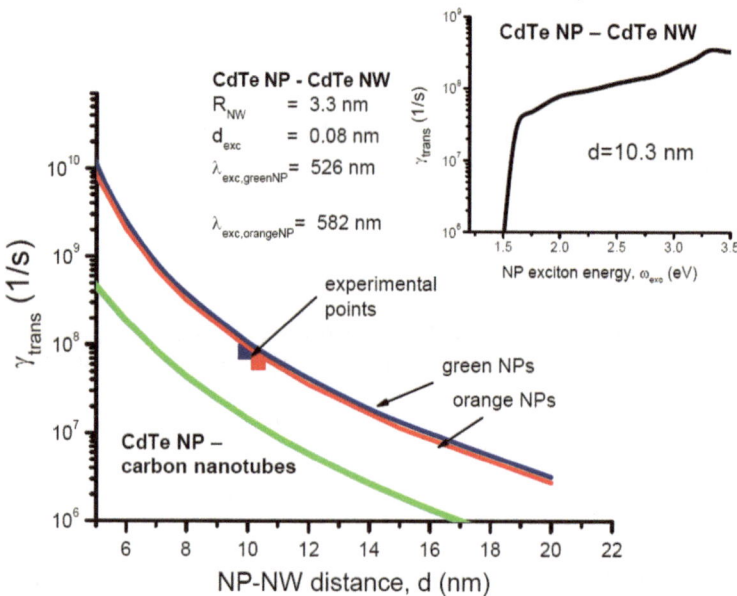

Fig. 2.8 Rates of NP-NW transfer of excitons as a function of the CdTe NP-NW separation and available experimental data from Ref. [10]. *Green line* shows the calculated rate for carbon nanotubes. *Inset* FRET rate for the NP-NW complex as a function of the exciton energy of a NP [Reprinted (abstract/excerpt/figure) with permission from Ref. [4] (Copyright 2008 by the American Physical Society)]

and 2.8). Figure 2.8 shows the dependence $\gamma_{trans}(\omega_{exc}, d = 10.3$ nm) as an inset. The function $\gamma_{trans}(\omega_{exc})$ reflects the frequency dispersion of the CdTe dielectric function, $\varepsilon_{NW} = \varepsilon_{CdTe}(\omega)$. The CdTe NPs and CNTs we can neglect the second term in (2.37) because of the strong depolarization effect for the electric field perpendicular to the CNT axis [11]. Therefore, equation takes the form:

$$\gamma_\alpha(\omega_{exc}) = a_\alpha \frac{2}{\hbar} \frac{R_{CNT}^2}{d^5} \left(\frac{ed_{exc}}{\varepsilon_{eff}}\right)^2 \frac{3\pi}{32} \, \mathrm{Im}\, \varepsilon_{CNT} \tag{2.38}$$

where ε_{CNT} is the "z" component of the dielectric constant averaged over a CNT volume and the averaged transfer rate is given by $\gamma_{trans}(\omega_{exc}) = (3/2)\gamma_0$. Note that NP $\to$ CNT transfer is slower compared to that for the NP–NW system due to the smaller effective cross-section of CNT compared to the CdTe NW. This section is reprinted (abstract/excerpt/figure) with permission from Ref. [4]. Copyright 2008 by the American Physical Society.

2.5   Summary

To summarize the FRET rates, Table 2.2 lists the transfer rates for the long distance asymptotic behavior in the dipole approximation. Table 2.2 illustrates the distance dependency for the FRET: (1) when the acceptor is an NP, FRET is inversely

Table 2.2 FRET rate summary for the long distance asymptotic limit

α-direction	Donor			Coefficients		Acceptor distance dependency
	NP	NW	QW			X → NP
x	$\varepsilon_{eff_D} = \frac{\varepsilon_{NP_D}+2\varepsilon_0}{3}$	$\varepsilon_{eff_D} = \frac{\varepsilon_{NW}+\varepsilon_0}{2}$	$\varepsilon_{eff_D} = \varepsilon_0$	$b_x = \frac{1}{3}$		$\gamma_{NP} \propto \frac{1}{d^6}$
y	$\varepsilon_{eff_D} = \frac{\varepsilon_{NP_D}+2\varepsilon_0}{3}$	$\varepsilon_{eff_D} = \varepsilon_0$	$\varepsilon_{eff_D} = \varepsilon_0$	$b_y = \frac{1}{3}$		
z	$\varepsilon_{eff_D} = \frac{\varepsilon_{NP_D}+2\varepsilon_0}{3}$	$\varepsilon_{eff_D} = \frac{\varepsilon_{NW}+\varepsilon_0}{2}$	$\varepsilon_{eff_D} = \varepsilon_0$	$b_z = \frac{4}{3}$		
	NP	NW	QW			X → NW
x	$\varepsilon_{eff_D} = \frac{\varepsilon_{NP_D}+2\varepsilon_0}{3}$	$\varepsilon_{eff_D} = \frac{\varepsilon_{NW}+\varepsilon_0}{2}$	$\varepsilon_{eff_D} = \varepsilon_0$	$a_x = 0$	$b_x = 1$	$\gamma_{NW} \propto \frac{1}{d^5}$
y	$\varepsilon_{eff_D} = \frac{\varepsilon_{NP_D}+2\varepsilon_0}{3}$	$\varepsilon_{eff_D} = \varepsilon_0$	$\varepsilon_{eff_D} = \varepsilon_0$	$a_y = \frac{9}{16}$	$b_y = \frac{15}{16}$	
z	$\varepsilon_{eff_D} = \frac{\varepsilon_{NP_D}+2\varepsilon_0}{3}$	$\varepsilon_{eff_D} = \frac{\varepsilon_{NW}+\varepsilon_0}{2}$	$\varepsilon_{eff_D} = \varepsilon_0$	$a_z = \frac{15}{16}$	$b_z = \frac{41}{16}$	
	NP	NW	QW			X → QW
x	$\varepsilon_{eff_D} = \frac{\varepsilon_{NP_D}+2\varepsilon_0}{3}$	$\varepsilon_{eff_D} = \frac{\varepsilon_{NW}+\varepsilon_0}{2}$	$\varepsilon_{eff_D} = \varepsilon_0$	$b_x = \frac{3}{16}$		$\gamma_{QW} \propto \frac{1}{d^4}$
y	$\varepsilon_{eff_D} = \frac{\varepsilon_{NP_D}+2\varepsilon_0}{3}$	$\varepsilon_{eff_D} = \varepsilon_0$	$\varepsilon_{eff_D} = \varepsilon_0$	$b_y = \frac{3}{16}$		
z	$\varepsilon_{eff_D} = \frac{\varepsilon_{NP_D}+2\varepsilon_0}{3}$	$\varepsilon_{eff_D} = \frac{\varepsilon_{NW}+\varepsilon_0}{2}$	$\varepsilon_{eff_D} = \varepsilon_0$	$b_z = \frac{3}{8}$		

This list shows the distance dependence of the FRET rate as a function of the acceptor's geometry. Also, this includes the effective dielectric constant effect, which is a function of the donor's geometry. X = NP, NW or QW [Reprinted (adapted) with permission from Ref. [1]. (Copyright 2013 American Chemical Society)] proportional to d^{-6} (2.14, 2.15, and 2.16); (2) when the acceptor is a NW, FRET is proportional to d^{-5} ((2.29) and (2.30)); and (3) when the acceptor is a QW, FRET is proportional to d^{-4} (2.35). This indicates that the donor dimensionality does not affect the functional distance dependency on the distance. To complete our analysis, Fig. 2.9a show the distance dependencies, given in Table 2.2. The energy transfer rates are presented as a function of d/d_0, where d_0 is the characteristic distance, which satisfies the asymptotic condition required for each case $(d \gg R_{NP,(NW)}, d \gg L_{QW})$. Figure 2.9b presents the energy transfer efficiency for the FRET as a function of d/d_0. In all cases, the FRET's distance dependency is given by the acceptor geometry and it is independent of the donor's geometry. Note

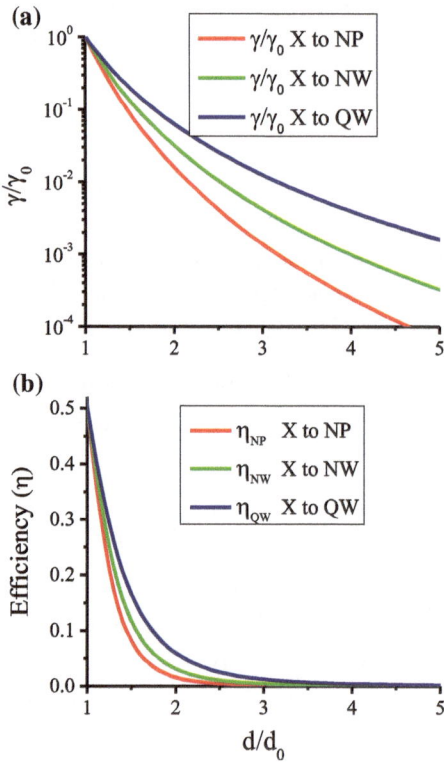

Fig. 2.9 a FRET rate distance dependency in the long distance asymptotic limit. Energy transfer rates are plotted as a function of d/d_0, where d_0 is the characteristic distance, which satisfies the asymptotic condition required for each case $(d \gg R_{NP,(NW)}, d \gg L_{QW})$. **b** Energy transfer efficiency for the FRET in the long distance asymptotic limit. Energy transfer efficiencies are plotted as a function of d/d_0. *Red line* shows the energy transfer efficiency for the D–A pair, when the acceptor is an NP. *Green line* depicts the energy transfer efficiency for the D–A pair when the acceptor is an NW. *Blue line* gives the energy transfer efficiency for the D–A pair when the acceptor is a QW. X = NP, NW, or QW [Reprinted (adapted) with permission from Ref. [1] (Copyright 2013 American Chemical Society)]

that the effective dielectric constant, however, depends only on the donor's geometry. Therefore, we can conclude that the FRET's distance dependency is dictated by the geometry of the acceptor nanostructure whereas the donor's contribution to the FRET appears through the effective dielectric constant. The dependencies given in Table 2.2 and Fig. 2.9 are important to understand FRET, and they are valid for the cases when the donor–donor and acceptor–acceptor separation distance is larger compared to the donor–acceptor separation distance. However, this condition is difficult to archieve experimentally and most of the experiments (in solid phase) are set using assembly of nanostructures. Therefore, it is crucial to understand FRET for the cases when the nanocrystals (NP and NW) are assembled into arrays (e.g., chains and films). This aspect is discussed in the following chapter.

References

1. P.L. Hernández-Martínez, A.O. Govorov, H.V. Demir, Generalized theory of Förster-type nonradiative energy transfer in nanostructures with mixed dimensionality. J. Phys. Chem. C **117**, 10203–10212 (2013)
2. A.O. Govorov, G.W. Bryant, W. Zhang, T. Skeini, J. Lee, N.A. Kotov, J.M. Slocik, R.R. Naik, Exciton-Plasmon interaction and hybrid excitons in semiconductor-metal nanoparticle assemblies. Nano Lett. **6**, 984–994 (2006)
3. E.D. Palik, *Handbook of Optical Constant of Solid* (Academic Press, New York, 1985)
4. P.L. Hernández-Martínez, A.O. Govorov, Exciton energy transfer between nanoparticles and nanowires. Phys. Rev. B **78**, 035314/1–035314/7 (2008)
5. S.K. Lyo, Exciton energy transfer between asymmetric quantum wires: effect of transfer to an array of wires. Phys. Rev. B **73**, 205322/1–205322/11 (2006)
6. S. Lu, A. Madhukar, Nonradiative resonant excitation transfer from nanocrystal quantum dots to adjacent quantum channels. Nano Lett. **7**, 3443–3451 (2007)
7. S.K. Lyo, Energy transfer from an electron-hole plasma layer to a quantum well in semiconductor structures. Phys. Rev. B **81**, 115303/1–115303/7 (2010)
8. D.L. Dexter, R.S. Knox, *Excitons* (Interscience Publishers, 1965)
9. T. Förster, in *Modern Quantum Chemistry*, ed. by O. Sinanoglu (Academic, New York, 1965)
10. J. Lee, A.O. Govorov, N.A. Kotov, Bioconjugated superstructures of CdTe nanowires and nanoparticles: multistep Cascade Förster resonance energy transfer and energy channeling. Nano Lett. **5**, 2063–2069 (2005)
11. C.D. Spataru, S. Ismail-beigi, X. Benedict, S.G. Louie, Appl. Phys. A **78**, 1129 (2004)

Chapter 3
Nonradiative Energy Transfer in Assembly of Nanostructures

This chapter is reprinted (adapted) with permission from Ref. [1]. Copyright 2014 American Chemical Society. Here, we present the theoretical framework of generalized Förster-type nonradiative energy transfer (FRET) between one-dimensional (1D), two-dimensional (2D), and three-dimensional (3D) assemblies of nanostructures consisting of mixed dimensions in confinement, namely, nanoparticles (NPs) and nanowires (NWs). Also, the modification of FRET mechanism with respect to the nanostructure serving as the donor versus the acceptor is discussed, focusing on the rate's distance dependency. Here, the combinations of X → 1D assembly of NPs, X → 2D assembly of NPs, X → 3D assembly of NPs, X → 1D assembly of NWs, and X → 2D assembly of NWs (where X is an NP, an NW, or a quantum well (QW) with the donor → acceptor (D → A) denoting the energy transfer directed from the donor to the acceptor) are specifically considered because they are important for practical applications. Furthermore, here we give a complete set of analytical expressions in the long distance approximation, for FRET in all of the cases mentioned above and derive generic expressions for the dimensionality involved to present a complete picture and unified understanding of FRET for nanostructure assemblies.

Let us first consider the energy transfer process from a single nanostructure (NP, NW, or QW) to assemblies of NPs and NWs. Specifically, we look at the following cases: (1) NP → 1D NP assembly (linear chain); (2) NP → 2D NP assembly (NPs layer or plane); (3) NP → 3D NP assembly; (4) NP → 1D NW assembly (plane); (5) NP → 2D NW assembly; (6) NW → 1D NP assembly; (7) NW → 2D NP assembly; (8) NW → 3D NP assembly; (9) NW → 1D NW assembly; (10) NW → 2D NW assembly; (11) QW → 1D NP assembly; (12) QW → 2D NP assembly; (13) QW → 3D NP assembly; (14) QW → 1D NW assembly; and (15) QW → 2D NW assembly. For all cases, an analytical expression for the

© The Author(s) 2017
P.L. Hernández Martínez et al., *Understanding and Modeling Förster-type Resonance Energy Transfer (FRET)*, Nanoscience and Nanotechnology, DOI 10.1007/978-981-10-1873-2_3

long distance approximation is given. We start this section with the macroscopic approach to the problem of dipole-dipole energy transfer.

The probability of an exciton transfer from the excited state of the donor nanostructure (donor) to the ground state of the acceptor nanostructure (acceptor) is given by the Fermi's Golden rule (3.1)

$$\gamma_{trans} = \frac{2}{\hbar} \left\{ \sum_f | \langle f_{exc}; 0_{exc} | \hat{V}_{int} | i_{exc}; 0_{exc} \rangle |^2 \delta (\hbar\omega_{exc} - \hbar\omega_f) \right\} \qquad (3.1)$$

where $|i_{exc}; 0_{exc}\rangle$ is the initial state with an exciton in the donor and zero exciton in the acceptor; $|f_{exc}; 0_{exc}\rangle$ is the final state with an exciton in the acceptor and zero exciton in the donor; $\hat{V}_{int}$ is the exciton Coulomb interaction operator; and $\hbar\omega_{exc}$ is the exciton's energy. As described in Chap. 5 from Understanding and Modeling Förster-type Resonance Energy Transfer (FRET) Vol. 1 (Refs. [2–4]), this expression can be simplified into

$$\gamma_{trans} = \frac{2}{\hbar} \text{Im} \left[\int dV \left(\frac{\varepsilon_A(\omega)}{4\pi} \right) \mathbf{E}_{in}(\mathbf{r}) \cdot \mathbf{E}_{in}^*(\mathbf{r}) \right] \qquad (3.2)$$

where the integration is taken over the acceptor volume, $\varepsilon_A(\omega)$ is the dielectric function of the acceptor, and $\mathbf{E}_{in}(\mathbf{r})$ includes the effective electric field created by an exciton at the donor side. The electric field is calculated with $\mathbf{E}(\mathbf{r}) = -\nabla\Phi(\mathbf{r})$ and the electric potential $\Phi(\mathbf{r})$ is given by

$$\Phi_\alpha(\mathbf{r}) = \left(\frac{ed_{exc}}{\varepsilon_{effD}} \right) \frac{(\mathbf{r} - \mathbf{r}_0) \cdot \hat{\boldsymbol{\alpha}}}{|\mathbf{r} - \mathbf{r}_0|^3} \qquad (3.3)$$

where ed_{exc} is the dipole moment of the exciton and ε_{effD} is the effective dielectric constant of the donor, which depends on the geometry and the exciton dipole direction, $\alpha = x, y, z$. Table 3.1 provides a summary for the donor dielectric constant as calculated for a single donor in Chap. 1 (Ref. [5]).

The average FRET rate (at room temperature) is calculated as

$$\gamma_{trans} = \frac{\gamma_{x,trans} + \gamma_{y,trans} + \gamma_{z,trans}}{3} \qquad (3.4)$$

Table 3.1 Effective dielectric constant expressions for the cases of NP, NW, and QW in the long distance approximation

α-direction	NP	NW	QW
x	$\varepsilon_{effD} = \frac{\varepsilon_{NP} + 2\varepsilon_0}{3}$	$\varepsilon_{effD} = \frac{\varepsilon_{NW} + \varepsilon_0}{2}$	$\varepsilon_{effD} = \varepsilon_0$
y	$\varepsilon_{effD} = \frac{\varepsilon_{NP} + 2\varepsilon_0}{3}$	$\varepsilon_{effD} = \varepsilon_0$	$\varepsilon_{effD} = \varepsilon_0$
z	$\varepsilon_{effD} = \frac{\varepsilon_{NP} + 2\varepsilon_0}{3}$	$\varepsilon_{effD} = \frac{\varepsilon_{NW} + \varepsilon_0}{2}$	$\varepsilon_{effD} = \varepsilon_0$

In this table the cylinder main axis is considered to be along the y-direction [Reprinted (adapted) with permission from Ref. [5] (Copyright 2013 American Chemical Society)]

where $\gamma_{\alpha,trans}$ is the transfer rate for the α-exciton ($\alpha = x, y, z$). In the following section the results obtained in Chap. 2 (Ref. [5]) are used to derive expressions for the assembly cases.

3.1 Energy Transfer Rates for Nanoparticle, Nanowire, or Quantum Well to 1D Nanoparticle Assembly

The FRET rate analytical equations are derived in the long distance approximation, when the donor is an NP, an NW, or a QW while the acceptor is a 1D NP assembly (linear chain) (Fig. 3.1). Assuming that the donor size is smaller than the separation distance between the D–A pair and using the long distance approximation, the energy transfer rate $\gamma_{\alpha,i}$ from the donor and the ith NP in the 1D NP assembly (chain) is given by

$$\gamma_{\alpha,i} = \frac{2}{\hbar} b_\alpha \left(\frac{ed_{exc}}{\varepsilon_{effD}}\right)^2 R_{NP_A}^3 \left|\frac{3\varepsilon_0}{\varepsilon_{NP_A}(\omega) + 2\varepsilon_0}\right|^2 \mathrm{Im}\, |\varepsilon_{NP_A}(\omega)| \frac{1}{(r^2 + y_i^2)^3} \qquad (3.5)$$

where $b_\alpha = \frac{1}{3}, \frac{1}{3}, \frac{4}{3}$ for $\alpha = x, y, z$, respectively; ed_{exc} is the exciton dipole moment; ε_{effD} is the effective dielectric constant for the exciton in the donor given in Table 3.1; ε_0 is the medium dielectric constant; R_{NP_A} and ε_{NP_A} are the acceptor NP radius and dielectric function, respectively; and r is the distance between the donor and linear NP chain (Fig. 3.1). The total transfer from the donor to all acceptor NPs in the chain is

$$\gamma_\alpha = \sum_i \gamma_{\alpha,i} = \frac{2}{\hbar} b_\alpha \left(\frac{ed_{exc}}{\varepsilon_{effD}}\right)^2 R_{NP_A}^3 \left|\frac{3\varepsilon_0}{\varepsilon_{NP_A}(\omega) + 2\varepsilon_0}\right|^2 \mathrm{Im}\, |\varepsilon_{NP_A}(\omega)| \sum_i \frac{1}{(r^2 + y_i^2)^3}$$

$$(3.6)$$

if the separation between NP is small and a linear density of particle λ_{NP} can be defined, then (3.6) can be rewritten as

$$\gamma_\alpha = \frac{2}{\hbar} b_\alpha \left(\frac{ed_{exc}}{\varepsilon_{effD}}\right)^2 R_{NP_A}^3 \left|\frac{3\varepsilon_0}{\varepsilon_{NP_A}(\omega) + 2\varepsilon_0}\right|^2 \mathrm{Im}\, |\varepsilon_{NP_A}(\omega)| \int_{-\infty}^{\infty} \frac{\lambda_{NP}}{(r^2 + y^2)^3} dy \qquad (3.7)$$

After integration, the expression boils down to

$$\gamma_\alpha = \frac{2}{\hbar} b_\alpha \left(\frac{ed_{exc}}{\varepsilon_{effD}}\right)^2 \left(\frac{3\pi R_{NP_A}^3}{8}\right) \frac{\lambda_{NP}}{d^5} (c_D)^5 \left|\frac{3\varepsilon_0}{\varepsilon_{NP_A}(\omega) + 2\varepsilon_0}\right|^2 \mathrm{Im}\, |\varepsilon_{NP_A}(\omega)| \qquad (3.8)$$

where d is the perpendicular distance between the donor and linear NP chain and c_D is a constant, which depends on the donor geometry; $c_D = 1$ for a NP, and $\cos(\theta_0)$ for a QW, and $(1 + \tan^2\theta_0 \sin^2\alpha)^{-1/2}$ for a NW. θ_0 is the angle between r and d as

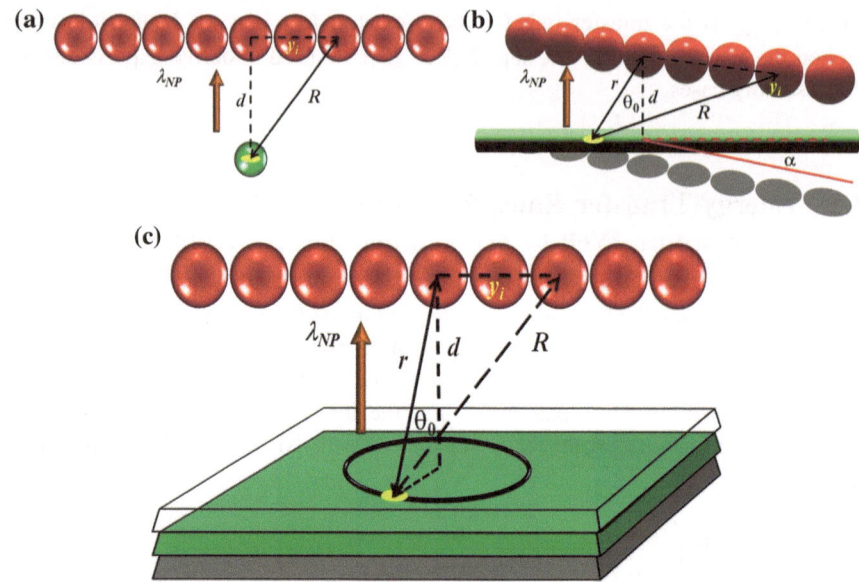

Fig. 3.1 Schematic for the energy transfer of **a** NP → 1D NP assembly, **b** NW → 1D NP assembly, and **c** QW → 1D NP assembly. *Orange arrows* show the energy transfer direction. *Yellow circles* represent an exciton in the α-direction. d is the separation distance. θ_0 is the azimuthal angle between d and r. α is the angle between NW axis and the NP array axis [Reprinted (adapted) with permission from Ref. [1] (Copyright 2014 American Chemical Society)]

shown in Fig. 3.1b, c. α is the angle between NW axis and the NP array axis (Fig. 3.1b). Note that the energy transfer rate distance dependency changes from $\gamma \propto d^{-6}$ to $\gamma \propto d^{-5}$. Furthermore, the FRET rate Eq. (3.8) strongly depends on the angle when the donor is a QW or NW.

3.2 Energy Transfer Rates for Nanoparticle, Nanowire, or Quantum Well to 2D Nanoparticle Assembly

We present simplified expressions for FRET rate in the long distance approximation when the donor is an NP, an NW, or a QW and the acceptor is a 2D NP assembly (plane) (Fig. 3.2). Similar to the previous case, we assume that the donor size is small compared to the D–A separation distance d. The energy transfer from a donor NP to the i, j-th acceptor NP in a 2D assembly is

$$\gamma_{\alpha,i,j} = \frac{2}{\hbar} b_\alpha \left(\frac{ed_{exc}}{\varepsilon_{effD}}\right)^2 R_{NP_A}^3 \left|\frac{3\varepsilon_0}{\varepsilon_{NP_A}(\omega) + 2\varepsilon_0}\right|^2 \mathrm{Im}\left|\varepsilon_{NP_A}(\omega)\right| \frac{1}{\left(d^2 + \rho_{i,j}^2\right)^3} \quad (3.9)$$

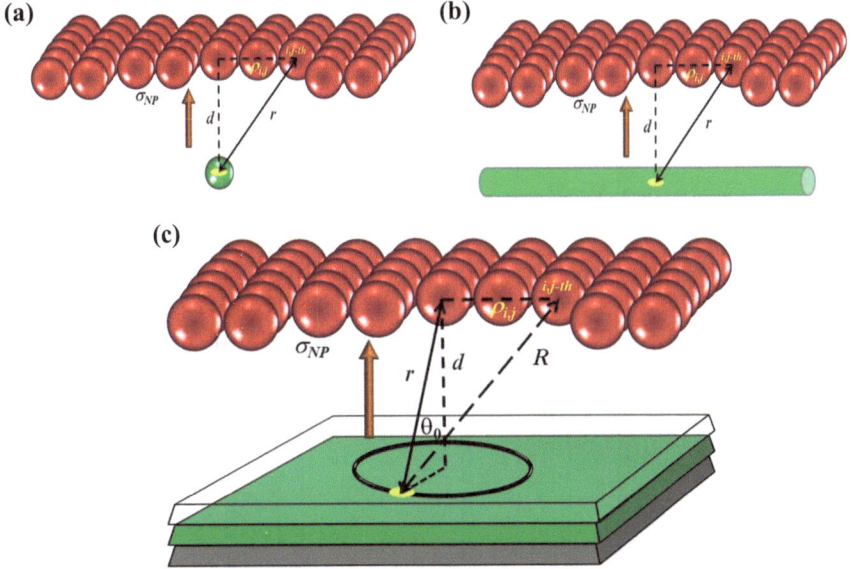

Fig. 3.2 Schematic for the energy transfer of **a** NP → 2D NP assembly, **b** NW → 2D NP assembly, and **c** QW → 2D NP assembly. *Orange arrows* denote the energy transfer direction. *Yellow circles* represent an exciton in the α-direction. d is the separation distance. θ_0 is the azimuthal angle between d and r [Reprinted (adapted) with permission from Ref. [1] (Copyright 2014 American Chemical Society)]

Thus, the total transfer rate is given by

$$\gamma_\alpha = \sum_{i,j} \gamma_{\alpha,i,j} = \frac{2}{\hbar} b_\alpha \left(\frac{ed_{exc}}{\varepsilon_{effD}}\right)^2 R_{NP_A}^3 \left|\frac{3\varepsilon_0}{\varepsilon_{NP_A}(\omega) + 2\varepsilon_0}\right|^2 \operatorname{Im}|\varepsilon_{NP_A}(\omega)| \sum_{i,j} \frac{1}{\left(d^2 + \rho_{i,j}^2\right)^3} \tag{3.10}$$

Assuming the separation between the acceptor NP is small and a surface density of particle σ_{NP} can be defined, (3.10) reduces to

$$\gamma_\alpha = \frac{2}{\hbar} b_\alpha \left(\frac{ed_{exc}}{\varepsilon_{effD}}\right)^2 R_{NP_A}^3 \left|\frac{3\varepsilon_0}{\varepsilon_{NP_A}(\omega) + 2\varepsilon_0}\right|^2 \operatorname{Im}|\varepsilon_{NP_A}(\omega)| \int_0^\infty \frac{2\pi\sigma_{NP}}{\left(d^2 + \rho^2\right)^3} \rho d\rho \tag{3.11}$$

The final equation for the transfer rate is

$$\gamma_\alpha = \frac{2}{\hbar} b_\alpha \left(\frac{ed_{exc}}{\varepsilon_{effD}}\right)^2 \left(\frac{\pi R_{NP_A}^3}{2}\right) \frac{\sigma_{NP}}{d^4} \left|\frac{3\varepsilon_0}{\varepsilon_{NP_A}(\omega) + 2\varepsilon_0}\right|^2 \operatorname{Im}|\varepsilon_{NP_A}(\omega)| \tag{3.12}$$

For this case, the distance dependency for the energy transfer rate is proportional to d^{-4}. This derivation is consistent with the previous studies reported in Refs. [6–8].

3.3 Energy Transfer Rates for Nanoparticle, Nanowire, or Quantum Well to 3D Nanoparticle Assembly

The FRET rate expression in the long distance approximation when the donor is an NP, an NW, or a QW while the acceptor is a 3D NP assembly is obtained (Fig. 3.3). In the same spirit as the previous cases, we assume that the donor size is small compared to the D–A separation distance d. The energy transfer from a donor NP to the i, j, k-th acceptor NP in a 3D assembly is

$$\gamma_{\alpha,i,j,k} = \frac{2}{\hbar} b_{\alpha} \left(\frac{ed_{exc}}{\varepsilon_{effD}} \right)^2 R_{NP_A}^3 \left| \frac{3\varepsilon_0}{\varepsilon_{NP_A}(\omega) + 2\varepsilon_0} \right|^2 \mathrm{Im} \left| \varepsilon_{NP_A}(\omega) \right| \frac{1}{\left(x_{ijk}^2 + y_{ijk}^2 + \left(z_{ijk} + d \right)^2 \right)^3}$$

(3.13)

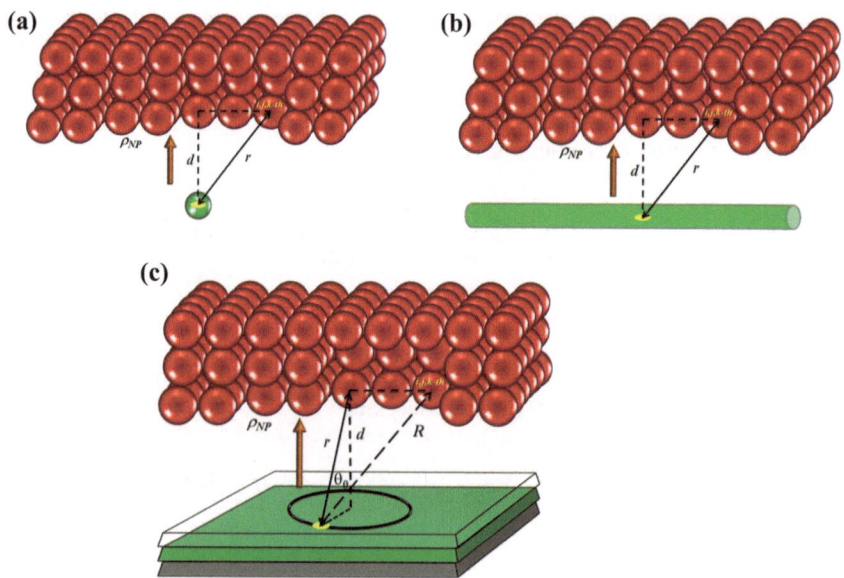

Fig. 3.3 Schematic for the energy transfer of **a** NP → 3D NP assembly, **b** NW → 3D NP assembly, and **c** QW → 3D NP assembly. *Orange arrows* denote the energy transfer direction. *Yellow circles* represent an exciton in the α-direction. d is the separation distance. θ_0 is the azimuthal angle between d and r. [Reprinted (adapted) with permission from Ref. [1] (Copyright 2014 American Chemical Society)]

Thus, the total transfer rate is given by

$$
\begin{aligned}
\gamma_\alpha &= \sum_{i,j,k} \gamma_{\alpha,i,j,k} \\
&= \frac{2}{\hbar} b_\alpha \left(\frac{ed_{exc}}{\varepsilon_{eff_D}}\right)^2 R_{NP_A}^3 \left|\frac{3\varepsilon_0}{\varepsilon_{NP_A}(\omega)+2\varepsilon_0}\right|^2 \operatorname{Im}|\varepsilon_{NP_A}(\omega)| \sum_{i,j} \frac{1}{\left(x_{ijk}^2 + y_{ijk}^2 + \left(z_{ijk}+d\right)^2\right)^3}
\end{aligned}
$$

(3.14)

Assuming the separation between the acceptor NPs is small and a volume particle density ρ_{NP} can be defined, (3.10) boils down to

$$
\gamma_\alpha = \frac{2}{\hbar} b_\alpha \left(\frac{ed_{exc}}{\varepsilon_{eff_D}}\right)^2 R_{NP_A}^3 \left|\frac{3\varepsilon_0}{\varepsilon_{NP_A}(\omega)+2\varepsilon_0}\right|^2 \operatorname{Im}|\varepsilon_{NP_A}(\omega)| \int_0^\infty \int_{-\infty}^\infty \int_{-\infty}^\infty \frac{\rho_{NP}}{\left(x^2+y^2+(z+d)^2\right)^3} dx\,dy\,dz
$$

(3.15)

The final equation for the transfer rate is obtained as

$$
\gamma_\alpha = \frac{2}{\hbar} b_\alpha \left(\frac{ed_{exc}}{\varepsilon_{eff_D}}\right)^2 \left(\frac{\pi R_{NP_A}^3}{6}\right) \frac{\rho_{NP}}{d^3} \left|\frac{3\varepsilon_0}{\varepsilon_{NP_A}(\omega)+2\varepsilon_0}\right|^2 \operatorname{Im}|\varepsilon_{NP_A}(\omega)|
$$

(3.16)

For this case, the distance dependency for the energy transfer rate is proportional to d^{-3}, similar to the bulk case [9].

3.4   Energy Transfer Rates for Nanoparticle, Nanowire, or Quantum Well to 1D Nanowire Assembly

We derive simplified expressions for FRET rate in the long distance approximation when the donor is an NP, an NW, or a QW with the acceptor being a 1D NW assembly (Fig. 3.4). Similar to the previous cases, we consider the energy transfer rate between the donor and the 1D assembly of NWs. In this case, the transfer rate to the i-th NW is

$$
\gamma_{\alpha,i} = \frac{2}{\hbar} \left(\frac{ed_{exc}}{\varepsilon_{eff_D}}\right)^2 \left(\frac{3\pi}{32}\right) R_{NW_A}^2 \left(a_\alpha + b_\alpha \left|\frac{2\varepsilon_0}{\varepsilon_{NW_A}(\omega)+\varepsilon_0}\right|^2\right) \operatorname{Im}|\varepsilon_{NW_A}(\omega)| \frac{1}{(d^2+y_i^2)^{\frac{5}{2}}}
$$

(3.17)

where $a_\alpha = 0, \frac{9}{16}, \frac{15}{16}$; $b_\alpha = 1, \frac{15}{16}, \frac{41}{16}$ for $\alpha = x, y, z$, respectively; ε_{eff_D} is the effective dielectric constant for the exciton in the donor NP given in Table 3.1; R_{NW_A} is the acceptor NW radius; and d is the distance between the donor and NW assembly (Fig. 3.4). The total transfer from the donor to all acceptor NWs in the chain is

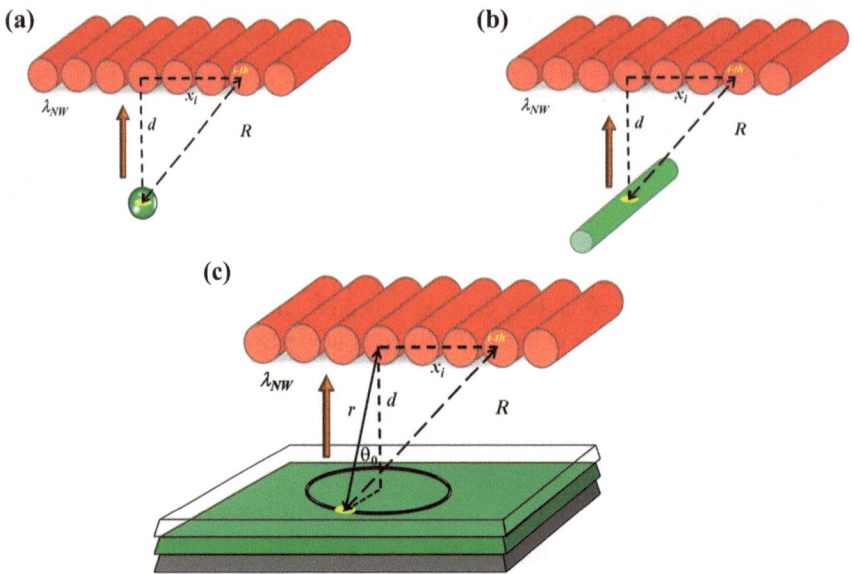

Fig. 3.4 Schematic for the energy transfer of **a** NP → 1D NW assembly, **b** NW → 1D NW assembly, and **c** QW → 1D NW assembly. *Orange arrows* show the energy transfer direction. *Yellow circles* represent an exciton in the α-direction. d is the separation distance. θ_0 is the azimuthal angle between d and r [Reprinted (adapted) with permission from Ref. [1] (Copyright 2014 American Chemical Society)]

$$\gamma_\alpha = \sum_i \gamma_{\alpha,i}$$

$$= \frac{2}{\hbar}\left(\frac{ed_{exc}}{\varepsilon_{effD}}\right)^2 \left(\frac{3\pi}{32}\right) R_{NW_A}^2 \left(a_\alpha + b_\alpha \left|\frac{2\varepsilon_0}{\varepsilon_{NW_A}(\omega)+\varepsilon_0}\right|^2\right) \operatorname{Im}\left|\varepsilon_{NW_A}(\omega)\right| \sum_i \frac{1}{(d^2+y_i^2)^{\frac{5}{2}}}$$

$$(3.18)$$

Under the assumption that the NWs are close to each other with a linear density λ_{NW},

$$\gamma_\alpha = \frac{2}{\hbar}\left(\frac{ed_{exc}}{\varepsilon_{effD}}\right)^2 \left(\frac{3\pi}{32}\right) R_{NW_A}^2 \left(a_\alpha + b_\alpha \left|\frac{2\varepsilon_0}{\varepsilon_{NW_A}(\omega)+\varepsilon_0}\right|^2\right) \operatorname{Im}\left|\varepsilon_{NW_A}(\omega)\right| \int_{-\infty}^{\infty} \frac{\lambda_{NW}}{(d^2+y^2)^{\frac{5}{2}}}\,dy$$

$$(3.19)$$

The final result is

$$\gamma_\alpha = \frac{2}{\hbar}\left(\frac{ed_{exc}}{\varepsilon_{effD}}\right)^2 \left(\frac{\pi R_{NW_A}^2}{8}\right)\left(\frac{\lambda_{NW}}{d^4}\right)\left(a_\alpha + b_\alpha \left|\frac{2\varepsilon_0}{\varepsilon_{NW_A}(\omega)+\varepsilon_0}\right|^2\right) \operatorname{Im}\left|\varepsilon_{NW_A}(\omega)\right|$$

$$(3.20)$$

It is observed that when the NWs are assembled with high density, the distance dependency for the transfer rate changes from d^{-5} to d^{-4}. A similar result can be found in Ref. [10] for the case of NW $\rightarrow$ 1D NW array.

3.5   Energy Transfer Rates for Nanoparticle, Nanowire, or Quantum Well to 2D Nanowire Assembly

The FRET rate expression in the long distance approximation when the donor is an NP, an NW, or a QW while the acceptor is a 2D NW assembly is presented (Fig. 3.5). In the same way as the previous cases, we consider the energy transfer rate between the donor and the 2D assembly of NWs. In this case, the transfer rate to the i, j-th NW is

$$\gamma_{\alpha,i,j} = \frac{2}{\hbar}\left(\frac{ed_{exc}}{\varepsilon_{effD}}\right)^2 \left(\frac{3\pi}{32}\right) R_{NW_A}^2 \left(a_\alpha + b_\alpha\left|\frac{2\varepsilon_0}{\varepsilon_{NW_A}(\omega) + \varepsilon_0}\right|^2\right) \mathrm{Im}\,|\varepsilon_{NW_A}(\omega)|\frac{1}{\left(y_{i,j}^2 + (d + z_{i,j})^2\right)^{\frac{5}{2}}}$$

(3.21)

The total transfer from the donor to all acceptor NWs in the array is

$$\gamma_\alpha = \sum_{i,j}\gamma_{\alpha,i,j} = \frac{2}{\hbar}\left(\frac{ed_{exc}}{\varepsilon_{effD}}\right)^2 \left(\frac{3\pi}{32}\right) R_{NW_A}^2 \left(a_\alpha + b_\alpha\left|\frac{2\varepsilon_0}{\varepsilon_{NW_A}(\omega) + \varepsilon_0}\right|^2\right) \mathrm{Im}\,|\varepsilon_{NW_A}(\omega)|\sum_i\frac{1}{\left(y_{i,j}^2 + (d + z_{i,j})^2\right)^{\frac{5}{2}}}$$

(3.22)

Under the assumption that the NWs are close to each other with a surface density σ_{NW},

$$\gamma_\alpha = \frac{2}{\hbar}\left(\frac{ed_{exc}}{\varepsilon_{effD}}\right)^2 \left(\frac{3\pi}{32}\right) R_{NW_A}^2 \left(a_\alpha + b_\alpha\left|\frac{2\varepsilon_0}{\varepsilon_{NW_A}(\omega) + \varepsilon_0}\right|^2\right) \mathrm{Im}\,|\varepsilon_{NW_A}(\omega)|\int_0^\infty\int_{-\infty}^\infty\frac{\sigma_{NW}}{\left(y^2 + (d + z)^2\right)^{\frac{5}{2}}}dydz$$

(3.23)

The final result is obtained as follows:

$$\gamma_\alpha = \frac{2}{\hbar}\left(\frac{ed_{exc}}{\varepsilon_{effD}}\right)^2 \left(\frac{\pi R_{NW_A}^2}{24}\right)\left(\frac{\sigma_{NW}}{d^3}\right)\left(a_\alpha + b_\alpha\left|\frac{2\varepsilon_0}{\varepsilon_{NW_A}(\omega) + \varepsilon_0}\right|^2\right)\mathrm{Im}\,|\varepsilon_{NW_A}(\omega)| \quad (3.24)$$

It worth mentioning that when the NWs are assembled into a high density 2D array, the distance dependency for the transfer rate changes from d^{-5} to d^{-3}. This behavior resembles the bulk case.

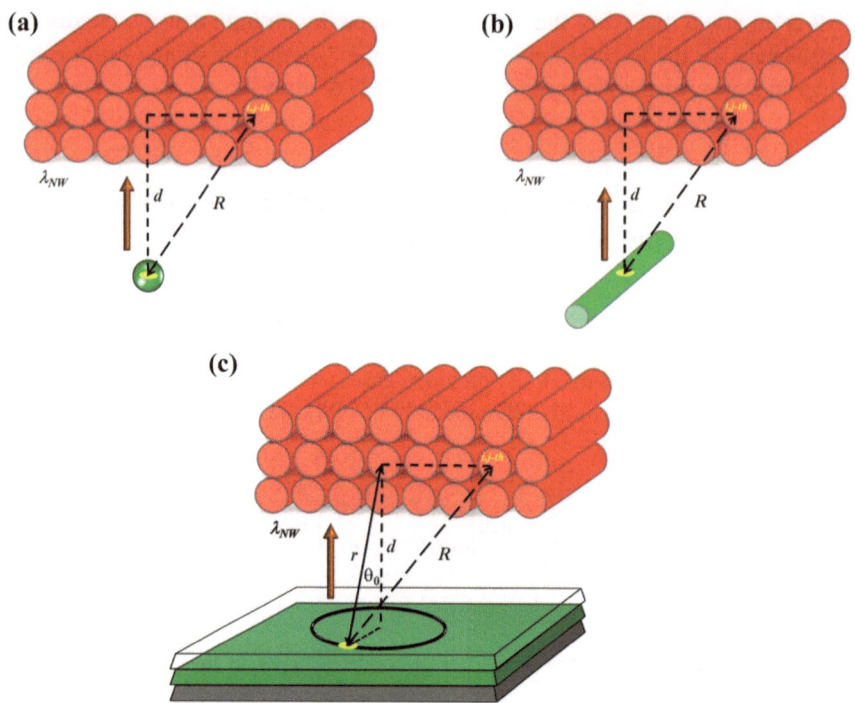

Fig. 3.5 Schematic for the energy transfer of **a** NP → 2D NW assembly, **b** NW → 2D NW assembly, and **c** QW → 2D NW assembly. *Orange arrows* show the energy transfer direction. *Yellow circles* represent an exciton in the α-direction. d is the separation distance. θ_0 is the azimuthal angle between d and r [Reprinted (adapted) with permission from Ref. [1] (Copyright 2014 American Chemical Society)]

3.6 Summary

A summary for all of the derived FRET rates is given in Table 3.2. Table 3.2 lists the transfer rates in the long distance asymptotic behavior in the dipole approximation for all combinations with mixed dimensionality (NP, NW, QW) in all possible arrayed architectures presented in this chapter (1D NP, 2D NP, 3D NP, 1D NW, 2D NW). This table illustrates the functional distance dependency for the FRET rates: (1) when the acceptor is an 1D NP assembly, the FRET rate is proportional to d^{-5} (3.8); (2) when the acceptor is an 2D NP assembly, the FRET rate is proportional to d^{-4} (3.12); when the acceptor is an 3D NP assembly, the FRET rate is proportional to d^{-3} (3.16); (4) when the acceptor is a 1D NW assembly, the FRET rate is proportional to d^{-4} (3.20); and when the acceptor is a 2D NW assembly, the FRET rate is proportional to d^{-3} (3.24). This suggests that the donor dimensionality (NP, NW, QW) does not affect the functional dependency on the distance. In all cases, the FRET rate distance dependence is given by the acceptor

Table 3.2 Generic distance dependency for the FRET rates, with equivalent cases of arrayed nanostructures in term of d dependence [Reprinted (adapted) with permission from Ref. [1] (Copyright 2014 American Chemical Society)]

Generic distance dependence	FRET Donor (D) $\rightarrow$ Acceptor (A)
$\gamma \propto \frac{1}{d^6}$	$X \rightarrow$
$\gamma \propto \frac{1}{d^5}$	$X \rightarrow \quad \equiv \quad X \rightarrow$
$\gamma \propto \frac{1}{d^4}$	$X \rightarrow \quad \equiv X \rightarrow \quad \equiv X \rightarrow$
$\gamma \propto \frac{1}{d^3}$	$X \rightarrow \quad \equiv X \rightarrow \quad \equiv X \rightarrow$

$X = $, , $\qquad$ d: separation distance between D and A $\equiv$: equivalent

geometry and acceptor array architecture and it is independent of the donor's geometry. Table 3.2 illustrates the FRET rate generic distance dependence with equivalent cases in term of d dependence. It is pointing out that the effective dielectric constant depends only on the donor's geometry. Therefore, we can conclude that the FRET's distance dependency is dictated by the confinement degree of the acceptor nanostructure and its stacked array dimensions whereas the donor's confinement affects the modification of effective dielectric constant.

References

1. P.L. Hernández-Martínez, A.O. Govorov, H.V. Demir, Förster-type nonradiative energy transfer for assemblies of arrayed nanostructures: confinement dimension vs. stacking dimension. J. Phys. Chem. C **118**(9), 4951–4958 (2014)
2. P.L. Hernández-Martínez, A.O. Govorov, Exciton energy transfer between nanoparticles and nanowires. Phys. Rev. B **78**, 035314/1–035314/7 (2008)
3. P.M. Platzman, P.A. Wolf, *Waves and interactions in solid state plasma* (Academic Press, New York, 1973)
4. A.O. Govorov, J. Lee, N.A. Kotov, Theory of plasmon-enhanced Förster energy transfer in optically excited semiconductor and metal nanoparticles. Phys. Rev. B **76**, 125308/1–125308/16 (2007)
5. P.L. Hernández-Martínez, A.O. Govorov, H.V. Demir, Generalized theory of Förster-Type nonradiative energy transfer in nanostructures with mixed dimensionality. J. Phys. Chem. C **117**, 10203–10212 (2013)

6. D.G. Kim, S. Okahara, M. Nakayama, Y.G. Shim, Experimental verification of Förster energy transfer between semiconductor quantum dots. Phys. Rev. B **78**, 153301/1–153301/4 (2008)
7. M. Lunz, A.L. Bradley, V.A. Gerard, S.J. Byrne, Y.K. Gun'ko, V. Lensyak, N. Gaponik et al., Concentration dependence of Förster resonant energy transfer between donor and acceptor nanocrystals quantum dots: Effects of donor-donor interactions. Phys. Rev. B **83**, 115423/1–115423/10 (2011)
8. X. Zhang, C.A. Marocico, M. Lunz, V.A. Gerard, Y.K. Gun'ko, V. Lensyak, N. Gaponik, A.S. Susha, A.L. Rogach, A.L. Bradley et al., Wavelength, concentration, and distance dependence of nonradiative energy transfer to a plane of gold nanoparticles. ACS Nano **6**, 9283–9290 (2012)
9. A.L. Rogach, T.A. Klar, J.M. Lupton, A. Meijerink, J. Feldmann, Energy transfer with semiconductor nanocrystals. J. Mater. Chem. **19**, 1208–1221 (2009)
10. S.K. Lyo, Exciton energy transfer between asymmetric quantum wires: Effect of transfer to an array of wires. Phys. Rev. B **73**, 205322/1–205322/11 (2006)

Appendix A

A.1 Useful Formalae

Förster-type Nonradiative Energy Transfer (FRET) Rate:

$$\gamma_{trans} = \frac{2}{\hbar} \text{Im} \left[\int dV \left(\frac{\varepsilon_A(\omega)}{4\pi} \right) \mathbf{E}_{in}(\mathbf{r}) \cdot \mathbf{E}_{in}^*(\mathbf{r}) \right] \tag{2.1}$$

FRET Rate in the Long Distance Approximation:

1. NP → NP

$$\gamma_{\alpha,trans} = \frac{2}{\hbar} b_\alpha \left(\frac{ed_{exc}}{\varepsilon_{effNP_D}} \right)^2 \frac{R_{NP_A}^3}{d^6} \left| \frac{3\varepsilon_0}{\varepsilon_{NP_A}(\omega_{exc}) + 2\varepsilon_0} \right|^2 \text{Im}[\varepsilon_{NP_A}(\omega_{exc})] \tag{2.14}$$

2. NW → NP

$$\gamma_{\alpha,trans} = \frac{2}{\hbar} b_\alpha \left(\frac{ed_{exc}}{\varepsilon_{effNW}} \right)^2 \frac{R_{NP_A}^3}{d^6} \cos^6(\theta_0) \left| \frac{3\varepsilon_0}{\varepsilon_{NP_A}(\omega_{exc}) + 2\varepsilon_0} \right|^2 \text{Im}[\varepsilon_{NP_A}(\omega_{exc})] \tag{2.15}$$

3. QW→NP

$$\gamma_{\alpha,trans} = \frac{2}{\hbar} b_\alpha \left(\frac{ed_{exc}}{\varepsilon_{effQW}} \right)^2 \frac{R_{NP_A}^3}{d^6} \cos^6(\theta_0) \left| \frac{3\varepsilon_0}{\varepsilon_{NP_A}(\omega_{exc}) + 2\varepsilon_0} \right|^2 \text{Im}[\varepsilon_{NP_A}(\omega_{exc})] \tag{2.16}$$

4. NP → NW

$$\gamma_{\alpha,trans} = \frac{2}{\hbar} \left(\frac{ed_{exc}}{\varepsilon_{effNP}} \right)^2 \left(\frac{3\pi}{32} \right) \frac{R_{NW_A}^2}{d^5} \left(a_\alpha + b_\alpha \left| \frac{2\varepsilon_0}{\varepsilon_{NW_A}(\omega_{exc}) + \varepsilon_0} \right|^2 \right) \text{Im}[\varepsilon_{NW_A}(\omega_{exc})] \tag{2.28}$$

© The Author(s) 2017
P.L. Hernández Martínez et al., *Understanding and Modeling Förster-type
Resonance Energy Transfer (FRET)*, Nanoscience and Nanotechnology,
DOI 10.1007/978-981-10-1873-2

5. NW → NW

$$\gamma_{\alpha,trans} = \frac{2}{\hbar}\left(\frac{ed_{exc}}{\varepsilon_{eff_{NW_D}}}\right)^2\left(\frac{3\pi}{32}\right)\frac{R_{NW_A}^2}{d^5}\left(a_\alpha + b_\alpha\left|\frac{2\varepsilon_0}{\varepsilon_{NW_A}(\omega_{exc})+\varepsilon_0}\right|^2\right)Im[\varepsilon_{NW_A}(\omega_{exc})]$$

(2.29)

6. QW → NW

$$\gamma_{\alpha,trans} = \frac{2}{\hbar}\left(\frac{ed_{exc}}{\varepsilon_{eff_{QW}}}\right)^2\left(\frac{3\pi}{32}\right)\frac{R_{NW_A}^2}{d^5}\cos^5(\theta_0)\left(a_\alpha + b_\alpha\left|\frac{2\varepsilon_0}{\varepsilon_{NW_A}(\omega_{exc})+\varepsilon_0}\right|^2\right)Im[\varepsilon_{NW_A}(\omega_{exc})]$$

(2.30)

7. NP, NW, QW → QW

$$\gamma_{\alpha,trans} = \frac{2}{\hbar}b_\alpha\left(\frac{ed_{exc}}{\varepsilon_{eff_D}}\right)^2\frac{1}{d^4}\left|\frac{2\varepsilon_0}{\varepsilon_{QW_A}+\varepsilon_0}\right|^2 Im[\varepsilon_{QW_A}(\omega_{exc})]$$

(2.35)

8. NP, NW, QW → 1D NP Assembly

$$\gamma_\alpha = \frac{2}{\hbar}b_\alpha\left(\frac{ed_{exc}}{\varepsilon_{eff_D}}\right)^2\left(\frac{3\pi R_{NP_A}^3}{8}\right)\frac{\lambda_{NP}}{d^5}(c_D)^5\left|\frac{3\varepsilon_0}{\varepsilon_{NP_A}(\omega)+2\varepsilon_0}\right|^2 Im|\varepsilon_{NP_A}(\omega)|$$

(3.8)

$c_D = 1, \cos(\theta_0)$ for NP and QW, respectively, and $\left(1 + \tan^2\theta_0\sin^2\alpha\right)^{-1/2}$ for a NW.

9. NP, NW, QW → 2D NP Assembly

$$\gamma_\alpha = \frac{2}{\hbar}b_\alpha\left(\frac{ed_{exc}}{\varepsilon_{eff_D}}\right)^2\left(\frac{\pi R_{NP_A}^3}{2}\right)\frac{\sigma_{NP}}{d^4}\left|\frac{3\varepsilon_0}{\varepsilon_{NP_A}(\omega)+2\varepsilon_0}\right|^2 Im|\varepsilon_{NP_A}(\omega)|$$

(3.12)

10. NP, NW, QW → 3D NP Assembly

$$\gamma_\alpha = \frac{2}{\hbar}b_\alpha\left(\frac{ed_{exc}}{\varepsilon_{eff_D}}\right)^2\left(\frac{\pi R_{NP_A}^3}{6}\right)\frac{\rho_{NP}}{d^3}\left|\frac{3\varepsilon_0}{\varepsilon_{NP_A}(\omega)+2\varepsilon_0}\right|^2 Im|\varepsilon_{NP_A}(\omega)|$$

(3.16)

11. NP, NW, QW → 1D NW Assembly

$$\gamma_\alpha = \frac{2}{\hbar}\left(\frac{ed_{exc}}{\varepsilon_{eff_D}}\right)^2\left(\frac{\pi R_{NW_A}^2}{8}\right)\left(\frac{\lambda_{NW}}{d^4}\right)\left(a_\alpha + b_\alpha\left|\frac{2\varepsilon_0}{\varepsilon_{NW_A}(\omega)+\varepsilon_0}\right|^2\right)Im|\varepsilon_{NW_A}(\omega)|$$

(3.20)

12. NP, NW, QW → 2D NW Assembly

$$\gamma_\alpha = \frac{2}{\hbar} \left(\frac{ed_{exc}}{\varepsilon_{eff_D}}\right)^2 \left(\frac{\pi R_{NW_A}^2}{24}\right) \left(\frac{\sigma_{NW}}{d^3}\right) \left(a_\alpha + b_\alpha \left|\frac{2\varepsilon_0}{\varepsilon_{NW_A}(\omega) + \varepsilon_0}\right|^2\right) \text{Im}|\varepsilon_{NW_A}(\omega)|$$

$$(3.24)$$

A.2 Effective Dielectric Constant

See Tables A.1 and A.2.

Table A.1 Effective dielectric constant expressions for NP, NW, and QW cases in the long distance approximation [Reprinted (adapted) with permission from Ref. [1] (Copyright 2013 American Chemical Society)]

α-direction	NP	NW	QW
x	$\varepsilon_{eff_D} = \frac{\varepsilon_{NP_D} + 2\varepsilon_0}{3}$	$\varepsilon_{eff_D} = \frac{\varepsilon_{NW} + \varepsilon_0}{2}$	$\varepsilon_{eff_D} = \varepsilon_0$
y	$\varepsilon_{eff_D} = \frac{\varepsilon_{NP_D} + 2\varepsilon_0}{3}$	$\varepsilon_{eff_D} = \varepsilon_0$	$\varepsilon_{eff_D} = \varepsilon_0$
z	$\varepsilon_{eff_D} = \frac{\varepsilon_{NP_D} + 2\varepsilon_0}{3}$	$\varepsilon_{eff_D} = \frac{\varepsilon_{NW} + \varepsilon_0}{2}$	$\varepsilon_{eff_D} = \varepsilon_0$

Table A.2 Generic distance dependency for the FRET rates, with equivalent cases of arrayed nanostructures in term of d dependence [Reprinted (adapted) with permission from Ref. [2] (Copyright 2014 American Chemical Society)]

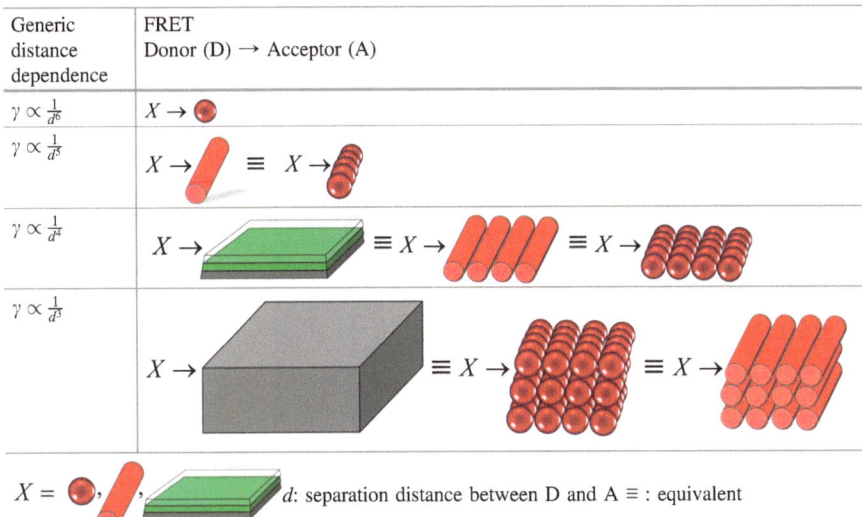

Generic distance dependence	FRET Donor (D) → Acceptor (A)
$\gamma \propto \frac{1}{d^6}$	
$\gamma \propto \frac{1}{d^5}$	
$\gamma \propto \frac{1}{d^4}$	
$\gamma \propto \frac{1}{d^3}$	

$X =$ ●, ▬, ▰ d: separation distance between D and A ≡ : equivalent

References

1. P.L. Hernández-Martínez, A.O. Govorov, H.V. Demir, Generalized theory of Förster-type nonradiative energy transfer in nanostructures with mixed dimensionality. J. Phys. Chem. C **117**, 10203–10212 (2013)
2. P.L. Hernández-Martínez, A.O. Govorov, H.V. Demir, Förster-type nonradiative energy transfer for assemblies of arrayed nanostructures: confinement dimension vs. stacking dimension. J. Phys. Chem. C **118**(9), 4951–4958 (2014)